Energy We Can Live With

Energy
We Can Live With

•

Approaches to Energy
That are Easy on the Earth and its People

Edited by
Daniel Wallace

A Collection of Articles from Rodale Publications
Organic Gardening and Farming • *Compost Science* • *Environment Action Bulletin*

Rodale Press, Inc.  Emmaus, Pennsylvania 18049

Library of Congress Cataloging in Publication Data

Main entry under title:

Energy we can live with.

"A collection of articles from Rodale
Publications, Organic gardening and farming,
Compost science, Environment action bulletin."
Includes index.
1. Power resources — Addresses, essays, lectures.
I. Wallace, Daniel.
TJ163.2.E494   333.7   76-28239
ISBN 0-87857-153-1

2   4   6   8   10   9   7   5   3

Printed in the United States of America on recycled paper

Contents

Preface

We haven't had many recent blackouts or brownouts, but we're still feeling the effects of the 1973 energy shortages. But perhaps there is a good side to the unpleasant memories and higher prices we're now paying for utilities, gasoline, food, and most other consumer items. At least they have made us more aware of how important it is for us to seek new energy sources and make the precious resources we now have last until the alternative sources become readily available.

Rodale Press was concerned about energy misuse and future shortages long before 1973. Our interests in the energy field really began with our interests in organic gardening and farming and natural foods, for organically oriented agriculture and minimal processing of food are low-energy alternatives to the extravagant high-energy methods of food production and preparation that have become characteristic of U.S. economics. Our interest in food soon spread into other energy-related areas. Now, home production, intermediate technology, land use, regionalism, recycling, and alternative sources of energy are familiar phrases in our books and magazine articles.

For several years now, every issue of *Organic Gardening and Farming, Environment Action Bulletin,* and *Compost Science* have focused on some aspect of energy, and *Energy We Can Live With* is an assemblage of the best of the recent articles on energy that have appeared in these publications.

If you look at the contents page, you'll probably be struck by the broad spectrum of ideas presented here. This was intentional; in fact, it was unavoidable. As we see it, there is no single solution to avoiding future energy crunches. Securing enough reliable, safe energy will most likely depend upon the use of several combined energy forms.

The first part of this book gathers together information on the more obvious alternate forms—the ones that have been making a lot of headlines lately: solar energy, methane gas, and wind power. It also explores some of the more provocative ideas for various energy sources, like getting alcohol from cellulose and burning it as fuel, turning sewage wastes into flammable "refuse derived fuel," and the potential of plain, old, human muscle power.

Because our interest in energy is so closely related to food production, and because food is so vital to us, both on a personal and a global survival level, the second section concerns itself with energy and agriculture. Here we look at the economics and efficiency of low-energy farming methods, the advantages of regional food markets, and even the feasibility of using draft horses again.

The last part puts a personal perspective on energy. It is here, perhaps, that alternate approaches to energy production and use are most real because none of them require advanced technology or years to make practical. Greenhouses that grow food in wintertime with the help of the sun are available now, as are wood stoves, compost piles, thermal underwear, and bicycles.

Introduction
Avoid the Energy Crunch

•

Robert Rodale

Editor's Note: Although written three years ago, the following article is as true today as it was then. American energy consumption is essentially unchanged, as surveys continue to show. United States energy consumption will more than double by the year 2000, and almost half of that energy will be wasted. But a good deal of this wasted energy can be saved if we take energy conservation measures more seriously, and even more if we follow up on some of the home production suggestions made in this article.

By now you have probably read plenty about the coming energy shortage. Newspapers, magazines, radio, and TV have all covered the story of the booming demand for fuel and energy of all kinds here, and the growing problems that surround the production of energy. There is little doubt in anyone's mind, after being exposed to all that information, that, as *Fortune* magazine says in a recent article, "The Energy Joyride Is Over."

For the sake of making sure that you know what we appear to be up against though, I will do some reviewing.

First, America is a country of energy drunkards. We comprise six percent of the world's population, yet we use about 35 percent of the world's energy production. And we keep *doubling* the amount of energy we use every 10 years, so you can see that the amount of energy we require for transportation, industrial production, heating, and air conditioning, is far more than that used by citizens of other countries.

Robert Rodale is the Publisher of ORGANIC GARDENING AND FARMING, ENVIRONMENT ACTION BULLETIN, *and* COMPOST SCIENCE.

Second, the amazing expansion of our energy needs is starting to strain our reserves of fuel. About 90 percent of the energy we use comes from fossil fuels—coal, oil and gas. (They are called fossil fuels because they are the fossilized remains of plants that for one period in the earth's history sprouted, bloomed, and died in fantastic abundance. In those primeval swamps, layer upon layer of giant ferns and similar plants grew and died without decaying into humus, but collected the way peat collects in bogs today. Later, ocean sediments covered those plant deposits, and they were compressed and processed naturally into the fuels we use today.)

Even though the amount of those fossilized fuels remaining under the surface of the earth boggles the imagination, our reserves are not limitless because of the rapid rate at which they are being used. Gas is scarce already. Oil is in critical supply in some areas of the globe, but abundant in other places. However, much of the oil that we use in the United States, especially in the populous East Coast area, comes from the Near East. The importation of that Arab oil strains our dollar reserves, and makes us politically dependent on the countries which supply the oil. It is not a healthy situation.

A third aspect of the energy problem has been created by our growing ecology consciousness and environmental awareness. Nuclear power plants, at one time a promised source of almost unlimited power, are coming along very slowly. People are afraid—rightly—of their accident potential and are worried about the problem of storing the atomic wastes, which must be kept safe from floods, earthquakes, and man-made disruption for thousands of years. Today, only a tiny fraction of our total power production comes from nuclear plants, and it appears that they won't supply great amounts of power for a long time to come.

Environmental problems also limit our fossil fuel reserves in several ways. For example, we have enough coal to last several hundred years, but people are objecting—again rightly—to the strip-mining process which devastates valuable lands. So although we have the coal, we may not be able to use it economically, because restoring strip-mined lands to their former quality is expensive. Coal is also dirty, creating air pollution problems even when coal-fired power plants are located far from cities. You probably have read about the large coal-fired power plants at Black Mesa, in Arizona, which are spreading clouds of pollutants over four states in order to produce enough electricity to supply the Los Angeles metropolitan area.

Pollution control itself requires energy, causing more rapid depletion of energy reserves. Cars equipped with emission controls get less mileage than older cars, because energy must be used to operate the air-purifying equipment. So our current push for cleaner air is causing a rapid increase in our already huge demand for oil.

Roughly, that's the energy picture, and it isn't a bright one. True, researchers around the globe are trying to tap totally new sources of energy. We read about breeder reactors that make their own atomic fuel, controlled fusion reactions that will release fantastic amounts of energy, plus a variety of other energy sources ranging from wind power, the tides, and geothermal plants that will tap the heat of the earth's molten core.

Don't expect any of those energy alternatives to make a big contribution to the massive power demands of the American economy in the near future, however. All of them are technologically sophisticated, and will require plenty of both time and good luck to be perfected. Remember how long it has taken for nuclear power to even begin to fulfill its promise of cheap energy. There's a good chance that the other new energy sources that are now promised will also prove to be disappointing. Our present high-energy civilization is based on simply lifting coal and oil from the earth and using it as concentrated power sources. No other kind of new energy promises to be nearly that easy to use.

The best evidence of all that a crunch is on the way is the fact that our leaders are now talking about saving energy as well as finding new sources of power. The Office of Emergency Preparedness, Executive Office of the President, published a staff study in October 1972, on "The Potential for Energy Conservation." It contains many useful ideas that could help us save some of the energy we have been wasting simply because power sources have been so cheap, at least up to now.

For example, many homes are not insulated nearly well enough. With good insulation in all homes, we could reduce the total energy needs of the United States by five percent, a remarkable figure. Super-convenient appliances are another source of energy drain. Frost-free refrigerators, for example, use 50 percent more power than conventional refrigerators. Did you realize that? Must you have that extra convenience, or can you defrost your refrigerator once in a while for the sake of saving electricity? Many air conditioners are wasteful of energy also. Eric Hirst and Robert Herendeen point out in an article in the October 28, 1972 *Saturday Review* that use of more efficient air conditioners, which are available, could save as much electricity as would be generated by three power plants as large as the mammoth one at Black Mesa in Arizona.

Transportation is revealed to be the biggest single energy consumer, requiring 25 percent of our total power needs. Jet planes are the most wasteful. Sending people and goods by air uses six times as much power as auto, truck, or rail transport. The authors of the *Saturday Review* article claim that eight percent of U.S. energy needs could be saved if more efficient methods of moving people and things were used.

Most of the commentaries about the energy problem fail to mention the extremely important role that cultivation of the soil plays in the whole power equation. In typical human fashion, people think only about the most visible and apparent power sources. They overlook the fact that green plants are the vital link that enables us to convert sun energy into food energy, and are far more important to our well-being than coal, oil, or gas.

Agriculture, in fact, is the original technique that man used to harness power to fuel an expanding economy. Plants have their roots in the almost inexhaustible mineral resources of the earth's crust, and their leaves are exposed to both the useful gases of the atmosphere (carbon dioxide is the gas plants need most) and to the rays of the sun. Through the remarkable process of photosynthesis, all these useful things are converted into carbohydrates, the food energy without which we could not survive. It's a remarkably simple and efficient process. And if it's handled right, photosynthesis is not only a non-polluting power source, but is constructively beneficial to the environment. The organic matter created by photosynthesis converts naturally into humus, improving the fertility of the soil. And the stored harvests of agriculture are just as much energy resources as are deposits of coal, or the charge of an electric battery.

The more organic a farming or gardening system, the more efficient is the energy yield of that enterprise. And by the same token, the more high-powered, technological, and agribusiness-oriented agriculture becomes, the less efficient is its yield of food energy—in relation to power consumed. Organic cultivation is more efficient because a greater proportion of its energy input comes from sun energy, which is free. Organic cultivation of the soil also is carried on to a large extent by human hands, aided with power equipment but not with the massive inputs of potent synthetic fertilizers and pesticides. Those chemical tools are made with the use of large amounts of fossil fuel energy. So when farmers and gardeners put synthetic materials on their land, they are in effect consuming the same kind of energy that powers automobiles and industrial plants, and air-conditions homes.

We don't yet know exactly how much more efficient organic growing is than conventional growing, from the point of view of energy consumption. That information can be computed by people who are experts in the study of energy. But we do know that conventional agriculture—at least as carried on in the United States—is no longer an energy-producing segment of our economy. Tractor fuel alone, without considering the energy needed to produce and transport synthetic fertilizers and pesticides, is about equal to the total energy yield of agriculture. That information comes from the excellent article "Farming

With Petroleum," by Michael J. Perelman in the October 1972 issue of *Environment.*

The main purpose of Perelman's article is to point out that American agriculture should not be cited for efficiency just because it can produce a lot of food with only a little human help. He cites as misleading the statement by Clifford Hardin, former Secretary of Agriculture, that "One man can take care of 60,000 to 75,000 chickens, 5,000 feedlot cattle, or 50 to 60 milk cows."

"After all," says Perelman, "no man alive can really feed 75,000 chickens by himself. In reality he is aided by many other men who make equipment and other necessities for raising chickens, even though some of them might never set foot on a farm." And all of those other men are consuming energy, and the products they create consume energy too when they are used on the farm. So even though our mechanized farms give the impression of tremendous efficiency, they are buying that efficiency with cheap fuel. *As the price of that fuel goes up, or as it becomes so scarce as to be unavailable, the inadequate energy yields of conventional farms will be exposed.*

And as the price of fuel goes up, more people are going to turn to gardening and small-scale farming as ways to produce needed food energy. There is plenty of evidence that human activity, especially when applied to making things grow in the soil, is a fantastically rewarding way to capture the energy of the sun and convert it into useful power. In his *Environment* article, Perelman cites the energy productivity of Chinese paddy culture of rice, which is so dependent on human labor that it really qualifies as a form of large-scale gardening. Chinese wet rice agriculture, says Perelman, can produce 53.5 British thermal units of energy for each Btu of human energy used in farming it. That means that for each unit of energy the farmer expands, he gets back over 50 in return.

By contrast, conventional agriculture returns only about one-fifth of a Btu in the form of food energy for each unit of fossil fuel that is expended in plowing, cultivating, harvesting, and storing crops. Looked at in that way, we can see that our vaunted agricultural productivity is extremely vulnerable to energy shortages, and is not nearly as rewarding per unit of energy expended as more primitive types of farming.

To sum up, there are two ways to avoid the power crunch organically. First, we must start conserving energy by using common sense. Better insulation of our homes is extremely important, and if you think of the word organic in a broad sense that is an organic technique. We must do more walking, especially around cities and towns, and less riding in automobiles. That certainly is organic.

And in deciding to do those things, we must not think that our little effort to conserve is meaningless when compared to the total picture of our energy-hungry civilization. Charles Lindbergh gave some advice on that point in an article "Lessons from the Primitive" in the November 1972 issue of *The Reader's Digest*. "Are we ready to moderate sufficiently our worship of bigness, power, speed, and affluence?" he asks. "Can we relearn the values of simplicity, tranquility, and balance? Are we bound to our technologies as an addict is bound to drugs, or can we follow a course of action based on human welfare? The answer must come from individuals like you and me, for civilization, with its governments and establishments, is shaped by the forces of human desire."

Finally and just as important, we must learn the energy lessons that agriculture and gardening can teach us. For I am sure that when the energy crunch does hit in full force, the home garden, the home storage shed for food, and the small orchard are going to be widely recognized as more important (and more liberating) energy sources than the ultra-high-technology generating plants now being worked on by researchers. We know that a good garden can yield plenty of storable energy with a moderate amount of work. By comparison, those fancy new power plants are just pipe dreams.

Part I

•

The Energy
Frontiers

Solar energy, wind power, and methane gas are not new in the energy arena. For as long as we've been growing food and burning wood we have been using solar energy. Indeed, the sun is the source of all our energy except that which comes from the splitting of the atom. The wind has powered sailing vessels and windmills for thousands of years, and methane gas, generated from organic waste, has been used as fuel in India and South Africa. What *is* new is our attitude towards these resources. We're beginning to realize their potential for heating, cooling, and the generation of electricity.

We're also taking a second look at some of the more obscure sources of energy. Geothermal power, which once seemed to have only regional potential, may be able to be tapped from many parts of the globe by pumping cold water into hot, dry underground regions and capturing the steam that would be produced to run turbines that would in turn generate electricity.

Cellulose wastes can be burned, but discoveries have been made in recent years that may make cellulose even more valuable. An enzyme has been isolated that can easily break down all forms of cellulose into glucose. Glucose can be used as a base for the production of alcohol, and alcohol can be used to stretch gasoline by replacing as much as 10 percent of any quantity of it.

We, in this country, think of bicycles more as pleasure vehicles than as equipment for doing work. Yet, in the People's Republic of China and in Taiwan, bicycles and other pedal-powered machinery like tricycle-trucks and foot-driven pumps and rail trolleys are quite common. They're convenient, easy to operate, at least as energy-efficient as their motor-powered counterparts, and could be useful in this and other highly advanced countries as well.

1

The Sun:
A Potential Powerhouse

•

Carol Stoner

Of all future sources of energy, solar energy seems to hold the most promise. It is clean, renewable, and available in good supply in almost all parts of the world. It is certainly the most abundant source of energy available to man: we are bombarded with thousands of times more energy every day than we could possibly use. The amount of energy that strikes just .5 percent of the land area of the United States is more than the total energy needs of the country projected to the year 2,000. (Allen L. Hammond, "Solar Energy: The Largest Resource," *Science*, September 22, 1972.)

But use of solar power is not just something for the future—we have always used it as a source of energy. As a matter of fact, all our present energy (except nuclear power) is derived from the sun. Wind power, wave power, and hydroelectric power are indirect sources of solar power, because the sun (and other celestial bodies) power the movement of both air and water bodies. All our food is just the sun's energy converted into plant tissue through the process of photosynthesis, and our fossil fuels are made from decayed vegetation that was once living and consuming the sun's rays.

All of these sources of energy are actually stored solar energy, and by relying on them for our entire energy supply, we are using only a minute fraction of all the sun's energy. A small fraction of one percent of the sun's energy is converted into plant tissue, and it is this tiny fraction that has produced all our fossil fuels, our food, and all other types of vegetation. (D. S. Halacy, *The Coming Age of Solar Energy*, New

Carol Stoner is the Executive Editor of the Rodale Press Book Division, and an Editor of ORGANIC GARDENING AND FARMING.

3

York: Harper & Row, 1973.) If we have gotten this much energy from the sun by making indirect use of it, just think of the energy we would have at our disposal if we could refine our means of collecting the sun's rays and use it more directly!

There are two fundamental ways we can use energy from the sun more efficiently than we have in the past. We can use it to generate electricity, and we can use it for climate control (heating and air-conditioning homes) and to heat water.

Using the sun's energy to produce electricity has attracted many researchers who believe that solar power, not nuclear power, is going to be our future source of electricity. Among these researchers is Dr. Karl Boer, Director of the Institute of Energy Conversion at the University of Delaware. He believes that solar cells can be built right into individual homes and other buildings to make electricity available at the site of installation. Dr. Boer has designed a house that is able to convert sunlight into electricity through solar cells built into its roof.

These solar cells, which are small, rectangular, chemically-treated wafers, are capable of collecting over 90 percent of the sun's rays. Even if they operate at only 10 percent efficiency, they could provide enough electricity to power all the lights and appliances in a modern home located in a sunny part of the country. The generation of electricity from both nuclear power and fossil fuels is incredibly inefficient: twice as much energy comes out of the process as waste heat than as electricity. With solar cells, however, there is no waste heat and no water, air, heat, or radioactivity pollution because no fuel is needed to produce electricity.

Among all the uses of solar energy, residential heating and cooling have the highest probability of success, according to the study done by the National Science Foundation and the National Aeronautics and Space Administration's Solar Energy Panel in late 1972. The panel predicts that by 1985, 10 percent of all homes and other buildings will use solar energy for as much as 80 percent of their heating and cooling requirements, and that by the year 2020, the proportion of new homes using solar energy will reach 85 percent.

Solar heating and air-conditioning systems operate on very basic principles. They are so simple and safe—because of the low temperatures they operate within—that they can be installed in small systems and maintained by unskilled homeowners.

The most important and most expensive part of the system is the solar collector. This is usually a flat, black panel that lies horizontally on the rooftop and absorbs almost all of the sun's rays falling on it. Water passing through pipes under this black panel heats up and passes into the storage area—usually a well-insulated water tank. The hot water in the tank can be pumped through pipes for bathing and washing. For space heating in winter, the hot water can be circulated through radiators, or fan blowers can transfer the heat from the water to living areas. In summer, the hot water can be used to run heat-operated air-conditioning units to cool the house. In most regions of the country a backup system based on conventional fuels would be needed for extended periods of bad weather.

Although such systems are being used on a limited basis today, much research is being done in the field and there are many success stories:

In the past 14 years, Dr. Harry Tomason has built himself three solar-heated houses in a suburb of Washington, D.C. His most recently built house works so well that the sun provides 70 percent of the heat needed in winter, and in the August heatwave last summer, the cooling system kept the house below 77 degrees F. without any difficulty.

Steve Baer, of Albuquerque, New Mexico, has himself a solar-heated house that was as simple to build as it is to maintain. Blackened drums of water lining the entire south side of the house absorb the sun's heat and transfer it to the living quarters.

Dr. Karl Boer, who uses solar cells to produce electricity in his experimental house at the University of Delaware, is also using the sun to heat and air-condition it. Special salts store heat collected by rooftop absorbers and transfer it to living spaces. In the summer, cool night air is used to cool these same salts, and the "cold" collected air-conditions the house during the daytime. Eighty percent of the heat and electricity will be supplied by the sun.

It has been estimated that if such systems were available now, solar heating and air-conditioning would be cheaper than electric heating and cooling in almost all parts of the United States, and would be competitive with gas and oil heating when these fuels double in price.

About the only drawback of solar heating and cooling is that the equipment necessary is not commercially available and may not be for a few years yet. Because of the initial large capital costs, industry may be hesitant about investing money for the manufacture of solar collectors, while homeowners may not readily accept this source of alternate heating and cooling. However, with the severe energy shortages and the rising costs of electricity and fossil fuels, homeowners may look forward to the day when such a relatively inexpensive and clean, reliable source of heating and cooling as solar energy is made available to them.

Editor's Note: In the last year a good number of private homes, industrial and public buildings, and schools using solar heating and cooling have been designed and are being built in many parts of the country. In

1976, the Federal government's Energy Research and Development Administration, under Program Opportunity Notice, gave out 36 grants totalling $8 million to schools and small businesses for the design and construction of solar heating units. Similar grants under the same program are expected to be awarded in 1977.

A Pond For Large-Scale Solar Heating and Grain Drying

•

Gene Logsdon

At first glance, the solar heat-collecting pond at Ohio's Agricultural Research and Development Center at Wooster looks like an Arab oilman's swimming pool. In fact, it looks that way at second glance, too. The pool measures 60 by 28 feet by 12 feet deep, and it's covered by a plastic bubble. Even on a sunny day in December (when I was there), the temperature under the bubble gets up to a warm 75 degrees F. or more, tempting one to dive in.

Yet, if anyone were so foolish as to do that, the resemblance to a swimming pool would change quickly to a similarity with the sea. The water is salty, and the deeper you go, the saltier it gets, reaching a 20-percent salt solution in the bottom six feet. Also unlike a conventional pool, the deeper you go, the warmer the water. In fact, on a very hot day in late summer, the water at the bottom of the pool might approach the boiling point, though now in December the pond is doing well to keep water on the lowest level at 100 degrees F.

The reason for the salt is to make the water denser, and therefore heavier at the bottom of the pool. The water was literally "laid" in the pool in layers, using a complex method worked out by Dr. Carl Nielsen of the Department of Physics at Ohio State. In the lower six feet, the water is 20 percent salt, with the salt content gradually decreasing to zero from the six-foot depth to the surface. In other words, the top layer is lighter than the layers below it—the key to how the water can hold and "store" the sun's heat. Only the bottom six feet of water are convective; when heated, that water will circulate only up to about the six-foot mark, rather than continue up to the top and dissipate the heat. So the

---

Gene Logsdon is Contributing Editor to ORGANIC GARDENING AND FARMING.

---

lighter top layer of water acts as an insulator, holding the heat in, while the water, being nearly transparent, allows the sun rays to go through. One meter of nonconvective water has an insulation value equivalent to six centimeters of Styrofoam.

The sunlight penetrates to the black liner on the bottom of the pool, which absorbs the heat and transfers it to the water. The ground under the pool stores heat, too, and can transfer it back to the pool in winter. The sides of the pool are insulated with Styrofoam. The bubble is composed of two layers of plastic with air between them for further insulation. And finally, to take advantage of every possible bit of warmth from the quicksilver sunlight of a winter day, the north wall of the structure is covered with a reflecting surface that increases the effective area of solar heat collection.

The long-range goal of the experiment is to see if such a solar pond can collect and store enough heat to warm a house of 2,000 square feet of living space comfortably through a northern winter. Presently, the Wooster experimenters intend to use the heat to warm a 1,000-square-foot greenhouse located right next to the pond. "Theoretically, we have enough heat for the greenhouse," says Philip Badger, who with fellow scientist Ted Short is in charge of the experiment. "But when you're blazing new trails, you learn not to make too many predictions. So far, so good."

Heat can be transferred from the bottom of the solar pond to the greenhouse in two ways: either by running fresh water through pipes in the bottom of the pond, where it would warm up and then be pumped back to the greenhouse; or by circulating the salty pond water itself back and forth between pond and greenhouse through the underground pipes. "At any rate, we intend to use a heat pump with the system," explains Badger. "That will improve performance considerably."

Meanwhile, at another location at the research station, solar heat is being put to use another way—to dry grain in quantity. This system uses air instead of water to harvest the sun's heat in two 10-mil vinyl plastic balloons that stretch 80 feet across the ground, and are 12 feet wide and 4 feet high. The covering top layer of a collector is clear plastic. A middle layer, inside, is tinted; and the bottom layer, right above the ground, opaque.

The layers are kept separated and supported by the air being pumped through them by a ½-hp centrifugal fan. Most of the air passes through the balloon below the tinted layer, but a small amount moves above it at a slower speed, which reduces heat loss through the top of the collector. (If you monkey around with convection and insulation and solar heat, you will appreciate what a slick trick that is. The collectors are made by Solar Energy Products Co. of Avon Lake, Ohio.)

Solar radiation is absorbed in the balloon primarily by the middle layer, and the heat harvested by the rapidly moving air beneath it. (Incidentally, I'm the only one who calls it a "balloon." The researchers all cringe at such "picturesque" writing. To them the thing is a plastic, air-supported solar collector. But balloon is much quicker.) The heated air is expelled out the end of the balloon next to the batch dryer, where it is picked up by the dryer fan and blown through the grain.

Does it work? Yes and no. The actual values of heat transferred to the dryer from the solar collector ranged from 60,000 Btu's per day to an impressive 800,000 per day, or the equivalent of one to nine gallons of propane per day. (Statistics are for the fall and winter of 1974/75; current year figures are not in yet.) That amount was enough to make soybean drying practical. Soybeans were dried at a bin depth of 36 inches from 16.1 percent moisture down to a safe storage level of 11 percent moisture in 54 hours.

However, drying shelled corn was not so successful. Very humid weather in the late October drying period caused grain to reabsorb moisture from the air faster than the solar dryer could take it out. Heat produced during January was not consistently high enough to dry the corn adequately, either.

But the scientists are learning as they go along. They found that when the collectors were oriented in an east-west direction, they caught more heat than when running north and south. So this year, both collectors run east-west. They also discovered that placing a layer of a Styrofoam insulation under the balloon resulted in better heating efficiency, even though the insulation blocked off much of the considerable heat exchange between the soil and the balloon that takes place during nighttime hours if the balloon is not insulated.

"Our preliminary data for the first year of trials indicates that solar energy collectors may be feasible for drying soybeans," cautiously reported George Meyer, Harold Keener, and Warren Roller at the 1975 meeting of the agricultural engineers. "But at this point, we certainly recommend a backup heating system to insure against spoilage losses in case of prolonged, unfavorable weather."

Although tests like these being conducted at Wooster are aimed at solving problems for larger commercial growers, they can be beneficial for gardeners and homesteaders, too. Even an experiment that doesn't pan out can be helpful or practical on smaller operations. For example, you or I may have a few bushels of beans to dry. Modest homemade plastic heat collectors and a small fan might be adapted to the purpose very well, using ideas generated by the larger experiments.

Or could a fish pond or swimming pool double as a solar heat collector? "You don't know how many people ask us that," says Badger with a grin. "Right now I don't see how that would be practical. Even a larger home swimming pool isn't big enough to heat a house or a larger greenhouse. And you'd have the salt in the water, don't forget. Fish couldn't live in our solar pond."

Nevertheless, the mad scientist in all of us knows there are definite possibilities for small projects. A swimming pool is large enough to heat a small greenhouse or a basement room or two. And in a solar pond, the surface water is neither very salty nor overly warm. There could be some way to find a solution to the salt solution if you want your solar pond to double as a fish pond or swimming pool.

Methane Gas From Algae

•

Carol Stoner

The sun's energy can be used in ways other than heating, air-conditioning, electricity production, and crop drying. It can be used through photosynthesis to produce algae that will in turn be used to produce methane gas.

Any kind of organic matter can be used to generate methane, by being fed into an anaerobic digestor to produce methane. Some scientists contend that it might be more practical to use these wastes to fertilize ponds in which algae is grown, and use this algae to produce the gas. Because algae grow so rapidly and so easily, more methane could be produced this way than by using the wastes directly.

Dr. William Oswald and Clarence W. Golueke at the University of California at Berkeley think that the algae-methane process has real possibilities. They have been growing algae in fertilized outdoor ponds, harvesting it, and dumping it into covered ponds where it ferments and produces methane in the process. Their work involves determining the efficiency of this system and finding the best use for the methane produced.

Their ponds have been averaging yields of algae in excess of 100 pounds per acre per day, with peak yields during the midsummer approaching 300 pounds per acre every day. By fertilizing the ponds with fresh chicken manure, they have been able to harvest as much as 450 pounds of algae per acre per day. Oswald and Golueke have figured that an annual yield of 25 tons per acre could yield 3,000,000 Btu per acre per year in the form of methane. (According to the Bureau of Mines, the United States consumed 72.1 quadrillion Btu's of energy in 1972.)

In an actual large-scale system, the methane produced could be used in place of fossil fuel, natural gas, or burned at the site in an electrical

11

generation plant. If the methane were the end product, as much as 65 percent of the algae energy could be recovered. If the methane were burned to generate electricity, only about 20 percent of the original energy in the algae would be recovered, but the waste heat and carbon produced in the process could be recycled. The waste heat could be used to stimulate algae growth and fermentation, thus improving the efficiency of the system.

Large areas of land and/or water would be required for these ponds, and this has produced much criticism of the system. However, Dr. Oswald points out that between 20 and 40 million square acres of land will be required for disposal of sewage and organic wastes anyway, and these present disposal methods add nothing to our gas reserves. By putting this same land to better use as dual-purpose algal production and waste-disposal systems, it would theoretically produce enough methane to replace one-fourth to one-half of the natural gas consumed nationally in all of last year. Dr. Oswald adds that there is also the possibility that this land could be used simultaneously for wetland parks, wildlife enhancement, and open space areas.

The cost of the gas produced from such systems is difficult to analyze because there are so many variables: climate, solar energy conversion efficiency, the costs of land and ponds, and the value of waste disposal. But Dr. Oswald believes that large-scale algal-to-methane systems in a suitable geographical location could approach competitiveness with current high-priced sources of gas for energy.

The NSF/NASA Solar Energy Panel, of which Dr. Oswald is a member, concluded in their December 1972 report, "An Assessment of Solar Energy as a National Energy Resource," that methane production from algae and floating water plants, although still in its infancy, could have a one percent impact on energy production by 1985 and a ten percent impact by 1995.

Methane: The Renewable Energy Source

•

Mark Schwartz

The production of methane gas has been proposed as a means of alleviating some of the crush of the "energy crisis." Methane, the chief component of most natural gas, also known as marsh gas, is produced in nature by the bacterial decay of vegetation and animal wastes in the absence of air—a process known as anaerobic digestion.

Urban and agricultural wastes commonly considered pollution and health hazards could be converted to methane, a relatively pollution-free and convenient fuel. Dr. John T. Pfeffer, a sanitary engineer at the University of Illinois, has said, "Methane from solid wastes could fill 11 percent of the United States' needs for natural gas." The National Science Foundation has granted $83,000 to the University of Illinois for additional research including construction of an experimental reactor. (*Chemical & Engineering News*; June 25, 1973)

H. L. Bohn has calculated that a feedlot producing 100,000 cattle annually produces 150,000 tons of dry organic waste. On the basis of 10 cubic feet per pound of waste, this could be converted by anaerobic digestion into at least three billion cubic feet of methane worth $510,000 to $990,000 and enough to supply the natural gas needs of 30,000 people at present rates of use (*Environment*, December 13, 1971). Approximately two billion tons of manure are produced annually in the United States, not to mention the untold sums of sewage sludge and garbage that could be used in a reactor for the production of methane.

"Home digesters" have been used for many years by some farmers in

Dr. Mark Schwartz is the Director of Research and Development at Rodale Press, and also the Chairman of the Soil and Health Foundation.

Germany and France. Feeding the digester poses no problem—almost all organic wastes are suitable for digestion including garbage, animal manures, human manure, grass clippings, leaves, crop residues, and paper.

Ram Bux Singh, director of a methane gas research laboratory at Ajitmal in northern India, states that it is possible to manufacture small, family-sized methane generators that can make every home and apartment in the United States at least semi-independent of outside power.

Singh has personally supervised the construction of 200 bio-gas plants, and with his staff, has dramatically simplified and miniaturized the design of methane production units. Singh recently reported, "It is now possible to install a small, prefab bio-gas plant in a home or apartment as easily as we now install a water heater. Plumbing is no more complicated." (*Senator Mike Gravel Newsletter,* May 29, 1973)

Methane's Big Test

•

Ray Wolf

It's taken a long time, but slowly the process of producing methane gas from animal manure has evolved from fantasy to reality in the world of public opinion. Many of the efforts in methane production are still on the small, "see there, it works," scale. But a group of Washington State engineers are currently finishing work on what will be the largest digester in America to test not only the principle, but the feasibility of methane production. The unit will handle 100,000 gallons of cow manure at a time.

In a methane digester, animal wastes are exposed to anaerobic bacteria (bacteria that thrive in an atmosphere containing no oxygen). The gas-producing process takes place in two stages. First, organic matter is liquified into volatile solids, of an acid nature. Second, methane-forming bacteria consume the volatile acids to produce the bio-gas that is captured in the digester. Bio-gas contains about 55–70 percent methane. Methane is the major constituent of natural gas, and burns virtually pollution free.

Being constructed at the State Reformatory Honor Farm at Monroe, Washington, the digester is designed to produce methane, improve the fertilizer qualities of manure, and serve as a demonstration and test project. The construction is being financed with a grant from the State of Washington's Department of Ecology.

The beginning of the methane project goes back a few years to when the dairy operation was plagued with odor and fly problems from manure storage. A new manure-management system was designed, with the

Ray Wolf is the Organic Living Editor for Organic Gardening and Farming, *and is a Contributing Editor to* Environment Action Bulletin *and* Compost Science.

cows in individual sawdust-bedded stalls in a concrete-floored, covered loafing shed with a forced water washout system. Daily, a tractor scrapes manure and washdown water to sumps where it can either be transferred to a storage lagoon or directly pumped to the fields to be applied with a manure gun. Although this new system solves sanitation problems, it is very wasteful of water, using almost seven parts water to wash out one part of manure. When the methane digester is working, the manure will be scraped from the stalls dry, and mixed with one part water to two parts manure before being put in the digester. The loafing barn will still be washed out, but with much less water.

Before construction began, it was necessary to determine if there was enough manure, how much gas would be produced, what the system would cost, how much space it would require, and how the gas could be used. To make these projections, a thorough feasibility study was prepared.

Of the many digester designs available, the group chose a method that somewhat parallels that used in sewage treatment plants. Known as a high-rate digester, the system requires the contents be constantly mixed to ensure that the biological process operates at or near capacity all the time.

To achieve this constant mixing, the digester uses a gas recirculation system wherein some of the gas produced by the digester is pumped to the bottom of the tank, then allowed to bubble to the top. The constant upward bubbling forms a hydraulic flow that mixes almost the entire tank's contents. The recirculation system runs continually.

With this system, undigested fibers that normally would form a scum blanket at the top of the digester, stopping gas production, are kept in circulation where they are subjected to further biological breakdown. The drawback to such a system is that it requires energy to run the pumps to recirculate the gas. Prototypes have shown that this energy loss is less than the increased gas production of such a system.

Manure is gradually added and removed from the digester daily. The average holding time in the digester is estimated at 17 days. This allows for maximum gas production, and lets the system use all manure produced, within a few days of production.

Animal manure is a good source of nitrogen for plants. However, the nitrogen in manure is in a form not readily usable by plants. This nitrogen must undergo a slow transformation into an available form. While this happens, either in a field, manure pile or storage lagoon, some of the nutrients leach out. This lost material causes water pollution problems when it finds its way to streams. In a stream it acts as a fertilizer for weeds and algae which then grow out of proportion and deplete the water's oxygen supply, often killing fish. This process is known as eutrophication.

In a digester, bacteria consumes much of the organic matter and some nitrogen is changed to an ammonia type, but in an organic form. The nitrogen then is concentrated in the digester slurry, because it is not used in the digestion process. This results in an increased nitrogen content of the digested slurry as opposed to raw manure. What that boils down to is that digested slurry has a higher nitrogen content than the raw manure; the nitrogen is in a form more usable by plants and without the pollution potential of raw manure. With increased nitrogen value, the farmer need not apply as much of the slurry to fields to gain the same results as raw manure.

Ecotope estimates the digester will produce about eight tons of nitrogen a year. This will be used to replace some chemical fertilizer currently bought every year. At the current price of $365 per ton for urea containing 46 percent nitrogen, a ton of pure nitrogen would cost $793. With a production of eight tons of nitrogen, the yearly fertilizer value of the digester is $6,344.

In addition to the value of the fertilizer, there is the estimated gas production of 12,000 cubic feet per day that will be used to heat the creamery boiler at the farm. The gas has a heat value of approximately 600 Btu's per cubic foot, or a total production of 7,200,000 Btu's per day. Some of the gas production (20 to 40 percent) will go to heat the digester to its optimum production temperature of 95 degrees F. Currently the farm pays 33 cents per gallon for fuel oil. At that rate, the estimated yearly value of the surplus methane would be $4,488. The combined estimated value of gas and fertilizer is $10,832 per year.

Up to this point the idea sounds just fine. However, the system will cost slightly more than $100,000. Some of this cost is for design and testing, but to the average farmer, such an outfit would still be in the vicinity of $70,000. This brings up the question of whether a farmer would be willing to spend that kind of money for a process that turns manure into manure (with some fertilizer improvement) and gives off some gas. Of course, the question is more complex than that, but for many people, that's what it boils down to.

The answer is that at this time the concept of methane generation is still being investigated, and its practicality is yet to be known. The project by Ecotope and the Honor Farm will go a long way towards answering many questions about methane generation on a farm.

Can Windmills Supply Farm Power?

•

Mark Schwartz

An answer to the energy crisis may be in the wind. This is not as far-fetched as it may sound—windmills have been used to generate small amounts of electrical energy since 1890.

The concept is appealing because it is so simple: wind pressure turns vanes or propellers attached to a shaft. The revolving shaft, through connections to various gears and mechanical or hydraulic couplings, spins the rotor of a generator. The generator creates an electrical current in transmission lines that are tapped for desired uses of electrical energy (*Environment 15*, No. 1, 1973).

The big question is whether windmills are technically and economically feasible. Based on experiences in a number of other countries, the answer appears to be a qualified "yes." However, the scale on which they are feasible is uncertain.

Thirty or forty years ago, wind-generated power played an active part in the electrification of many rural homesteads all across the country. Then the power companies came along and demonstrated that they could provide more power at a lower cost by burning coal in huge plants and distributing power on a network of wires to individual users. But as we realize now, the cost was in the consumption of an irreplaceable natural resource, thousands of acres of earth mangled by strip-mining and the pollution of the air we breathe and water we drink.

How do we go about harnessing this clean, self-renewing power source? Dr. William E. Heronemus, Professor of Engineering at the University of Massachusetts, has proposed the construction of 300,000 wind turbines in the Great Plains, spaced as closely as one per square mile. Each 850-foot tower would carry 20 turbines consisting of a two-bladed, 50-foot propeller. The Great Plains network could provide the equivalent of a nuclear power plant with the capacity of 189,000 megawatts (million watts). The wind-powered systems actually would have considerably larger total installed capacity in terms of potential electricity generating power, but the potential would never be met over a year's time since wind speed variations would make it impossible to keep the generators turning at maximum output.

The comparison with nuclear power plant capacity was worked out mathematically by Dr. Heronemus to provide a more realistic measure of the electrical production that could be anticipated from wind power. The 189,000 megawatt capacity of the wind network would be significant, nonetheless, since the total capacity of electricity generating plants in the United States in 1970 was only an estimated 360,000 megawatts (*Environment 15*, No. 1, 1973).

Dr. Wendell Hewson, Chairman of the Atmospheric Sciences Department at Oregon State University, is studying the use of wind power along the Oregon coast. He emphasizes that wind power would be only a supplement to existing power sources, not a replacement for them. But at certain places and at certain times of the year, it could supply 10 to 15 percent of our national power needs.

It is difficult to determine the cost of wind power at this stage of development. However, Dr. Hewson points out, "with more research and technology, the possibilities are greater for lower costs when one considers that the cost of our conventional power is expected to increase, whereas, the cost of wind power after research and installation would be expected to decline."

The critics of the use of wind power are quick to point out its inefficiencies and raise the old bugaboo: "What happens when the wind dies down?" A team of scientists in Geneva, Switzerland, has developed a new type of windmill. Walter Schoenball who heads the group said, "The new windmill makes use of 70 to 80 percent of the available wind, while the old-fashioned windmills trapped an average of 10 percent of the wind and a maximum of 30 percent." He also commented that "weather statistics show that in most parts of the world there are only three days a month, on average, without any wind at all. And in any case the power generated on windy days can be stored."

The use of wind power is enticing—and I believe feasible. In fact, as the realization of our energy problems becomes more apparent, other forms of energy besides wind power, such as solar and geothermal energy—which were also at one time said to be impractical and extremely costly—are beginning to look appealing.

Storing Wind-Generated Electricity

A group of young Texas researchers has developed a plan to generate electric power without using fossil fuels such as oil, gas, or coal.

Gray Company Enterprises, Inc. (GCOE) says the easily obtainable alternative is hydrogen, and has issued a considerable amount of information to back up its claim. The research is still in the early stages, and much work has yet to be done. But the initial results appear promising: Hydrogen may be the storage unit for energy generated by the wind.

The process itself seems simple: The wind drives a windmill which, connected to a generator, produces electricity. This electricity is used to electrolyze water—break it into its component parts, hydrogen and oxygen.

The hydrogen, in turn, would be used in much the same way as natural gas is currently used—with the additional advantages of being able to power automobiles and industrial equipment, says the company.

The basic advantage to this system is that it is highly efficient and pollution-free. The product of combustion would be pure water, and little, if anything, more. There is no noise, says the company, and no parts to wear out.

Even on a theoretical level, however, there are problems to be overcome. "Undoubtedly hydrogen is a hazardous material," says a GCOE report, "and must be handled with all due precautions. If it is handled properly, however, in equipment designed to ensure its safety, anyone should be able to use it without hazard."

Because the gas is extremely light, the company says, it leaves a vented space quickly rather than accumulating as natural gas and gasoline fumes tend to do.

Many people, the report continues, tend to fear hydrogen in what is called the "Hindenburg syndrome"—a fear derived from the famous 1937 airship disaster. "Spectacular as it was, however," the company adds, "that fire was almost over within two minutes, and of the 97 persons on board, 62 survived.

If strict codes are developed for hydrogen use and transport, as they have been for natural gas, there should be little danger involved, the company asserts.

The issue of pollution-free energy must, of course, be weighed against public safety. And, since the project is still in its early stages, new and undiscovered problems may yet arise. But the idea is intriguing and, with proper support, investigation and development, could become a factor in solving the crisis of our rapidly depleting resources. It appears, at the least, worthy of consideration by the Energy Research and Development Administration.

Riding the Wind

There ain't no racin' clippers now,
nor never will be again,
And most o' the ships are gone by now,
the same as most o' the men . . .
                                    —Rudyard Kipling

When Britain's balladeer penned his eulogy to the tall ships, graceful clippers still sliced the seas. Four-masters carried tea from China and grain from Australia 'round the Horn to European ports. But by the 1930s, the forests of masts that once shot skyward along the world's waterfronts were thinned considerably, and the few remaining deepwater square riggers were among the first victims of World War II. Today they are only memories, survived by a handful of museum pieces, tourist-carrying cruise ships and cadet training vessels.

But those saddened by the passing of the age of sail need mourn no more. The tall ships are coming back, ready to smash deepwater freight-carrying records, all on clean, free wind power, according to an article by James McCawley in the November 1971 issue of *Rudder*.

Even as you read this, Berlin's Stahlform steel mills are fabricating masts for these new clippers, vessels called Dyna-Ships, whose main method of propulsion will be the wind and whose main means of support will be carrying cargo at a cost its designers hope will drive shippers to their adding machines and shipowners directly into the arms of the Hamburg syndicate that thought of the idea in the first place.

There are no real Dyna-Ships at this moment, only models and computer readouts from tank and wind tunnel tests that show the new

22

"square-riggers" to be 60 percent more efficient than all other systems in use today. In simulated runs across the Atlantic, the Dyna-Ship design averaged 12 to 16 knots and hit top speeds of 20 knots. It also has the ability to sail fast in wind-whipped seas that would slow down conventional freighters.

According to its designers, the design combines lower operating costs with more cargo space, gained by replacing huge diesel power plants and fuel tanks with a small auxiliary unit. Repeated tests have convinced these men that their ships could compete with diesels for 90 percent of the bulk shipping market, using only one-twentieth of the fuel burned by today's cargo carriers. That is an added bonus that should make both ecologists and those concerned about the "energy crisis" happy.

But how can a sailing ship suddenly become so efficient? If you're thinking in terms of rigging and wooden masts, wipe that picture from your mind. The Dyna-Ship is a four-master, but it resembles the old clippers about as much as an F-4 Phantom resembles a World War I Sopwith Camel. The Dyna-Ship carries no standing or running rigging, sets sails by computer and the canvas, which is stored on drums inside the masts, is run out on stainless steel yardarms curved for the highest-possible aerodynamic efficiency.

Still, the projected 1973 launch date for the first Dyna-Ship might never have come about without Wilhelm Proelss, now 70, who began to think about the Dyna-Ship principle even before World War II. He recalls the unusual circumstances that sparked his research into efficient sailing systems.

As a young employee, Proelss was sitting in on a company meeting which, like many meetings, was rapidly degenerating into a slumber party. In an attempt to stay awake, he looked out over Hamburg's bustling harbor, but only one sight caught his attention. "I could see out the window a sight I thought very sad," he says. "A tug was towing a broken-down four-masted barque for disposal in the scrapyard. It bothered me so much to see that.

"Ten thousand years of human progress thrown on a scrap heap. 'Must this be?' I thought. 'Are they really so inefficient?' That was the beginning."

Proelss carried his idea around for 21 years—he resolved to design an efficient windjammer every year, but something always seemed to get in his way—until 1956, when he began his calculations based on the knowledge of aerodynamics he acquired during his work with fueling systems at Germany's eight major airports. Sixteen months later he had a small wooden model of the Dyna-Ship, based on his preliminary designs and calculations.

At this point, Proelss needed testing facilities and many more detailed figures about the sea, winds and ship dynamics, figures that were available in only one place, the *Institut fur Schiffbau Technischen* (the School for Naval Architecture) at the University of Hamburg.

The institute is equipped with everything a naval architect needs. It has wind tunnels, tank testing facilities, machine shops, computers, and, most important, money. The annual budget for research and development is more than a half-million marks, or about $150,000.

Proelss knew this, and one morning, without an appointment, he walked to the school where, fortunately, he was received by *Direktor* Gunther Kempf. Proelss at the outset must have felt like a Columbus trying to convince people that the world was round, for Kempf was skeptical as the engineer unpacked his portfolio. But skepticism gave way to interest, then complete absorption as the hours passed.

The ship Proelss had designed carried masts that soared 200 feet into the air, but there were neither lines nor shrouds and little work for the crew. The yardarms were made of curved steel fitted with tracks on which the sails would roll out from the center of the mast. What Proelss had designed, in effect, was a continuous airfoil from the top of each mast to the bottom, the angle of which was set by turning the mast hydraulically from the bridge. One man could handle every sail on the Dyna-Ship by pushing a few buttons.

He also came prepared with diagrams and energy conversion charts, showing the extremes of stress in all conditions of winds and waves. These Proelss had computed from the daily weather charts of the German Hydrographic Institute, and they formed the basis of the dozens of theoretical Atlantic crossings he had plotted.

It made an impressive show. When it was over, Kempf, and later George Weinblum of the Hamburg Research Council, pledged their support.

Quietly, a research team began work on the project, but it did not take long before the word spread elsewhere. It was then that Alan Villiers, a British author, sea captain, and advocate of the square-rigger, pledged his support after studying Proelss' design. Said Villiers afterward, "He has produced a wholly modern concept of the wind-driven ship at sea."

The major work on the Dyna-Ship at the School for Naval Architecture took six years. It was financed by the Hamburg Research Council and directed by Hans Thieme, an associate professor at the school. Thieme had earlier written a promotion-winning paper on sailing ship efficiencies, unaware at the time that it would become the bible of the Proelss Dyna-Ship supporters.

At the *Institut*, researchers built scale models of Proelss' design and refined them through wind tunnel tests. Next, they put the design to a challenging test: They made models of other typical sailing vessels from a modern seven-meter racing yacht to the *Preussen*, Germany's famed five-masted square-rigged freighter, and tested the designs in the wind tunnel against the Dyna-Ship. After seeing the results of the tests, Proelss said, "It was amazing. I would not have believed it myself, but we could actually see how clean the design was, how no part of the rig interfered with the flow of wind (the research team had used nitrogen "smoke" in the wind tunnel so they could observe the efficiency of various designs) past any other part . . . and yet how virtually all of the available wind energy was transformed into thrust for the ship."

Finally, the Dyna-Ship team built a full-scale section of mast and sail. During thousands of tests, the sail performed perfectly, in all wind strengths, from all angles of attack.

It was 10 years from the time Proelss had knocked on Kempf's door when the school released the first of several lengthy reports on the Dyna-Ship, estimating that such a vessel could return to its owner a 30 percent greater profit on investment because of its lower operating and construction costs and greater hold capacity.

The Dyna-Ship had proved itself to a dedicated group of men of science, but fleet owners in several countries were unconvinced. A return to sails seemed a step backward.

"Sailing ships?" one German shipowner asked in dismay. "My grandfather lost his shirt on them. They've been dead since 1900."

In England, the story was the same, Villiers recalls. "I took the idea to some enterprising young shippers I knew in Liverpool. They liked what they saw and were very interested, but the timing was all wrong. Freight shippers were switching to container ships, and my friends, in order to survive, were putting all of their extra capital into containerization."

Thus, although the Dyna-Ship had the backing of the naval institute, no one was willing to spend a single mark or pound on sails.

Aside from their interest in containerization, shippers raised questions about whether the wind could be counted on to keep a ship on a reasonable schedule. It was this lack of predictability that finally sank the old square-riggers, they said.

According to the institute, that assumption is not based on fact. Those areas of the ocean where there is no wind at any given time are very small. With its auxiliary motors moving the vessel 200 miles a day, even the worst equatorial calms can be traversed in just two days. Furthermore, the institute points out, in these days of advanced weather forecasting and weather satellites, it would be easy for a Dyna-Ship captain to plot the most favorable route just as airline pilots routinely alter their courses to avoid turbulence and other unfavorable weather.

From computer readouts of several hundred Atlantic crossings, the institute showed that winds were sufficient 72 percent of the time to keep Dyna-Ships moving on course at speeds up to 20 knots. Average speeds ranged from 12 to 15 knots. But best of all, fuel consumption was computed to be only 5 percent of that consumed by conventional freighters of equal size.

Entering the 1970s, however, there was still not one single shipper who would back the Proelss square-rigger. One company that had earlier gone through several talking stages went bankrupt, and the idea seemed dead. Proelss felt that his life's work had come to naught.

Then help came from an unexpected source: environmentalists with sizable bank accounts.

Rudolph Zirn, a Bavarian lawyer who runs one of the biggest law firms in Austria and Germany, commissioned the first Dyna-Ship, a five million dollar cruise ship with a tentative launch date of 1973. Zirn's consuming interests are in ships and sailing and preserving the remaining wild areas of the earth. He wants to take small tour groups to visit these areas to create an interest in saving them. And he wants to take them on a vessel that will not pollute the treasures these places hold, nor the waters around them.

Friederich Beutelrock of Lubeck, Germany, is another customer. Friederich and his brother are heirs to a shipping business built up by their father. At the moment they have seven conventional ships under construction, but number eight will be a Dyna-Ship.

Enter also George Kiskaddon, a prominent San Francisco shipper who heads the Oceanic Society, a group of environmentalists whose goal it is to rid the oceans of pollution. Kiskaddon is convinced that Dyna-Freighters can help clean up the seas, and he is talking about buying a Dyna-Ship the society can use for oceanographic studies. He is also trying to line up freight contracts for the Beutelrocks.

David Brower, president of Friends of the Earth (FOE), is also interested. Brower sees a government-funded Dyna-Ship project as a suitable substitute for the SST which FOE helped to stop.

Meanwhile, no decision has been made on where the hull of Zirn's Dyna-Ship will be built, nor is he sure he is going to find bankers to share the investment risk with him. He may take the whole gamble himself.

Nevertheless, there seems to be enough momentum now to envision Dyna-Ships of some sort in oceanic travel during this decade. The tall ships will once again thrust acres of sail into the sky. Like ghosts of the clippers, they will sail in silence. No men in the rigging, no shouted commands from the deck, no singing as sheets are rolled in and anchors raised. If you listen close all you might hear is the clicking of computers, the sound of the wind and the crying of seagulls in the sky—a clear, blue sky once again.

Editor's Note: Although the proposal to build a full-size Dyna-Ship has not yet been realized, active research continues. Dyna-Ship rigging has been tested on light sailboats on Priest Lake, Idaho, and other tests have been conducted on the San Francisco Bay. Interest has further grown in the Dyna-Ships due to the failure of nuclear power to convert easily to ship propulsion.

Waste As An Energy Resource

•

Daniel Wallace

During the past few years, concern over the world's limited energy resources has become the dominant factor in shaping thoughts about future economic development. Correspondingly, attention has been drawn to the demonstrated properties of municipal waste as an energy resource. Anaerobic digestion has produced methane gas, and sludge has been proposed in many places, and introduced in just a few as a replacement for chemical fertilizers which are quickly becoming very expensive in money and energy.

Myriad experiments and actual programs have proven sludge to be most valuable as a fertilizer to produce crops. The city of Denver has been depositing 500 to 600 wet tons of sludge, (equivalent to 100 ton dry), each day on land growing wheat for livestock. Farmers in Lima, Ohio, have been using sludge on cornfields, and Chicago has begun to market composted municipal solids as the organic fertilizer, "Nu-Earth." In Braunschweig, Germany, waste has been used as farm fertilizer since 1900, and in Edinburgh, England, sewage fields were established to provide fertilizer as early as 1843. China has been using "night soil," composted human waste, on their farmlands for centuries.

The success of these programs in terms of resource conservation has been remarkable. One farmer was estimated by the regional office of the EPA at Bowling Green, Ohio, to have saved $40 an acre in chemical fertilizer costs with a seven ton per day application of dried sludge. Agricultural experts agree that with applications of from five to ten dry tons per acre, no additional commercial fertilizer except potassium would be needed. Russell Train, EPA administrator, said that the use of

Daniel Wallace is an Associate Editor of the Rodale Press Book Division.

feedlot waste and municipal sludge could supply 6.5 percent of the nation's nitrogen, thereby saving 400 million annually in chemical fertilizers. If other major cities of the United States used sludge as fertilizer, total annual production would exceed nine million tons, enough to bring present fertilizer prices down, with much compost left over for exportation.

Knowing all this, it's hard to understand why the country is slow to accept the waste-to-fertilizer concept. One genuine fear is the presence of heavy metals in sludge which could be absorbed by plants fertilized through land application. Heavy metals vary in amounts in different communities, depending upon industrial disposal of metals in sewage. Denver's wheat crops are used as livestock feed, thus filtering out any metals through another medium. Other farm usage of sludge application depends upon a low, safe level of heavy metals. In actuality all heavy metal concerns can be banished by careful control of industrial wastes, and, the heavy metal can be reclaimed and used again, another resource-saving device.

Heavy metal as an impediment to sludge fertilization is really a scientific concern, not well-known to the general public. A major difficulty in advancing the use of sludge is the public's tendency to identify sludge as "raw sewage," in spite of the success of Denver, Lima, Chicago, and the centuries-long use of it in Europe and the Far East. Sludge application to land is not contaminating, is cheaper than chemical fertilizers, and is more productive, too. However, convincing the average citizen of this is a difficulty not easily surmounted.

Another use of municipal waste might pave the way—burning refuse and sludge as fuel. At the Third Annual Composting and Waste Recycling Conference in 1973, Dr. Clarence Golueke of the University of California included in his remarks that experiments in St. Louis were taking place which used composted sludge as fuel. Dr. Golueke saw these experiments as valuable not only in offering a supplement to coal, but also in leading to an expanded market for compost. Composting plants might be built for fuel, and eventually diversify by burning low-grade waste while returning high-grade composted waste to the soil. The public would eventually become acclimated to a complete use of waste in this fashion.

In fact, St. Louis processes 325 tons of garbage each day, recovering metals for recycling, while burning enough organic matter to make electricity to power the equivalent of 25,000 homes. The city authorities project a growth of up to 6,000 tons per day by 1977.

Other cities have followed suit. Ames, Iowa, put into operation in 1975 one of the most comprehensive resource recovery systems in the nation, reclaiming steel and aluminum from waste and reprocessing the rest into a supplemental fuel for pulverized coal. Baltimore burns garbage to make steam, handling 1,000 tons of refuse each day, and San Diego uses organic waste to make fuel oil. Ronald D. Kinsey, recycling consultant for the city of St. Paul, predicts that approximately 15 trillion Btu's of energy, the equivalent of three million barrels of oil, will be generated annually through garbage-to-energy systems in the upper Midwest by 1985.

The benefits from fossil fuel and mineral resources saved by these systems are more than creditable. A steel ton made of recycled materials saves 74 percent of the energy used in processing virgin ore and 40 percent of the freshwater, eliminates 76 percent of the water pollutants, and nullifies 2.7 tons of mining waste that would have been strewn upon the landscape.

The urban economics of garbage-to-energy systems prove to be desirable also. St. Louis paid roughly five dollars per ton to incinerate or landfill refuse. With the energy recycling system St. Louis now pays only a net two dollars per ton, after profits from the reclaimed metals and fuel.

The city of Coventry, England, has made an arrangement with an existing manufacturing plant to provide them with fuel from their waste heat. The Stokes manufacturing plant had an average fuel bill as of 1973 of $1,165,000 per year. The Coventry incinerator, burning refuse, provides the same amount of energy for $745,600, saving the plant an estimated $419,400 while generating considerable extra revenue for the city's use.

These recycling programs profit from reclaimed metals and energy directly derived from burned refuse, but they fulfill only partially the waste to energy ideal expressed by Dr. Golueke.

Odessa, Texas, has devised a program called LandTill which is based upon three components: recovery of metals, recovery of energy from refuse and wastewaters, and water conservation from wastewater and precipitation. Odessa's main goal is to improve its soil's capability to retain water, due to a poor yearly rainfall. In order to effect this goal, the city is placing shredded organic refuse and wastewater on the land to improve its retentive qualities. The project is estimated to cost Odessa $50,000 per year, with a return of one million dollars or more per year in new revenue. But the most noteworthy part of this program is the fact that sewage will also be used to irrigate the land, thus completing the extremely efficient waste-to-fertilizer facet of the energy recycling concept. With the success of programs such as this, the vision of a complete waste recycling system for the world, including waste as fertilizer, may be in the not-too-distant future.

Combining Alternate Energy Systems

•

Sandra Fulton Eccli

In this age of specialization, we are accustomed to looking for the "single" solution—the one cure-all—to solve our problems, and most Americans tend to look for such a cure-all to end our energy woes. But it should be understood that any single source of alternative energy— whether it be sun, wind, or what-have-you—will not be sufficient to supply the complete energy needs of a homestead, farm, or small community. We must think in terms of combining these sources to work in harmony with each other and with the natural environment.

Why should one source—say, solar power—not be adequate to supply all the power needs of a household or a community? For one thing, the supply of most of these new resources is intermittent. And each energy source comes to us in a different form. Solar energy arrives as heat, and it is most efficient to use it as heat. Wind and water come in mechanical form. Fuels from organic matter (methane gas, wood, and alcohol represent three of these) are generally more portable and versatile, but not that plentiful. Since no single source will, in all likelihood, be adequate, a combined energy system will have to consist of a diverse and harmonious matrix of the various energy sources, their storage, interconnections, final energy use, and, where possible, recycling back into the system.

Although very new in concept, a few combined systems have already been built, or are in the process of being built. For instance, near Al-

Sandra Fulton Eccli is the former Coeditor of ALTERNATIVE SOURCES OF ENERGY, and edited ALTERNATIVE SOURCES OF ENERGY: PRACTICAL TECHNOLOGY AND PHILOSOPHY FOR A DECENTRALIZED SOCIETY.

buquerque, New Mexico, Robert and Eileen Reines, with associates, have built a system that combines solar heating with electric power derived from the wind. On the outskirts of the city of London, England, Bruce Haggart and Graham Caine have completed the initial construction of their "Street Farmhouse"—formerly called "Eco-House"—combining solar heat with agricultural and methane production and a wind-electricity system, showing that combination systems can be achieved even in a large city.

Ms. A. N. Wilson of Martinsburg, West Virginia, has achieved the combination of solar power, wind generation of electricity, and anaerobic composter to produce, if not methane, high-quality compost. Her house, completed in the fall of 1973, contains approximately 1,400 square feet of living space (including four bedrooms) and 350 square feet of mechanical equipment space. Together, the wind system, with battery storage, and the solar complex supply 80 percent of all heating and power needed by this large, comfortable house.

Perhaps the most ambitious project to date is the one conceived by Richard Blazej of Newfane, Vermont. "Grassy Brook Village," presently undergoing construction of Phase I, will consist of 20 residential houses—a small village, actually a condominium. But what a condominium! It is designed to make the minimum impact on the natural environment and provide life-support systems that derive their energy from natural and non-polluting sources to the greatest possible degree. A further goal is to assure the resident-owners a high degree of financial and quality control.

Located on a 43-acre tract of woodland near Brookline, Vermont, Grassy Brook, when completed, will feature solar energy for house heating and hot water, generation of electricity from wind power, on-site handling of wastes in a system (still experimental) designed to operate in several recycling and pollution-free operations, and, hopefully, the production of methane for fuel.

Life in Grassy Brook Village will be anything but "primitive." Architect Robert F. Shannon and Project Engineer Fred Dubin (well-known for his work in solar energy systems), have designed a complex in which the standard of living will most likely be higher than in the average American suburb, because the design includes parks, long vistas for enjoying scenery, and minimal cost for utilities. Roof gardens provide beauty—and additional insulation. The complex will make maximum use of insulation, reducing heat demand of each house in the complex to approximately 7,500 Btu's per day.

Developer Blazej is organizing Grassy Brook Village as a condominium to provide the necessary legal structure for the use of common facilities, and to provide an opportunity for each family to own an ultramodern, energy-saving dwelling. Of all the plans for utilizing combined systems for the benefit of people, Grassy Brook seems to be the most ambitious thus far. If the experimental waste treatment systems work well, Grassy Brook will be, within the next couple of years, the *first* village-scale application of mixed alternate energy systems—solar, wind, waste treatment, and methane fuel production.

A Recycling System for Single-Family Farms and Villages

•

C. G. Golueke and W. J. Oswald

In the last few years, considerable attention has been directed to developing a home living unit that offers a maximum of self-sustenance while it is in complete harmony with its external environment. The chief concern is to put a minimum burden on both the environment and available resources (especially sources of energy).

The ideal goal is to create a series of "closed" systems in which residues are directly recycled into the individual living unit that generated them.

During our research on the use of algae in the reclaiming of nutrients and water from municipal and agricultural wastewaters, and on the development of photosynthetic life-support systems for extraterrestrial applications, we designed a self-contained living system. It consists of an algal regenerative system which has the added advantage of providing for the use of solar energy. Thus it combines addition with preservation, as external solar energy is brought into the system to augment the energy that is conserved within the system.

Bringing in outside energy is essential because no system can operate without a net loss of energy. In our system, solar energy keeps everything running, just as solar energy keeps everything running on earth. Our system has, in miniature, the features of the living part of the earth—photosynthesis for crop production; aerobic and anaerobic bacterial decomposition for the carbon and nitrogen cycles; the recy-

Dr. Golueke is Research Biologist and Lecturer, Sanitary Research Laboratory, University of Calif.
Prof. Oswald is in the Department of Engineering and School of Public Health, University of Calif.

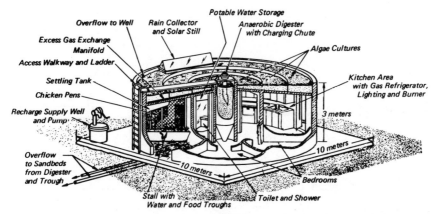

Figure 1. Schematic Diagram of a Dwelling Unit for a Family of four and their Livestock which Incorporates a Microbiological Recycle System for Water, Nutrients, and Energy in a Convenient and Hygienic Environment.

cling of water; plus the use of the chemical energy of methane. This system is beyond the planning stage because its components have been demonstrated in laboratory and pilot-scale studies to be technologically feasible both individually and integrally.

Besides the people and animals, the principal components of the system are: 1. an anaerobic digester; 2. a series of algal growth chambers; 3. a sedimentation chamber; 4. sand beds; 5. a solar still; 6. a gas exchanger.

Because it is combined with a home that requires gas for cooking, the anaerobic digester is covered to allow combustible gas to accumulate under a pressure sufficient to force it from the collector to the stove. Excess gas rich in methane (55–65 percent) is conveyed from the gas dome through conduits into the residence where it is also used for heating. Digested solids are periodically drawn from the digester for use as soil-conditioner or fertilizer in the growth of vegetables on a nearby soil plot.

At its minimum practical size, the algal regenerative system will provide waste disposal and nutrient recycling for four humans, one cow and 50 chickens. This size is arbitrarily considered as the most elementary that is operable. The bases upon which the size of the components of this single-family unit were estimated are described subsequently. But within certain size limits (as yet to be determined), design data proven satisfactory for the single-family unit can be directly expanded to fit larger populations.

A diagrammatic sketch of a typical family unit of the dimensions given is presented in Figure 1. The operation of the system involves the charging of all manure, urine, wasted food, night soil, and cleanup water into the digester shortly after they are produced, or at least once a

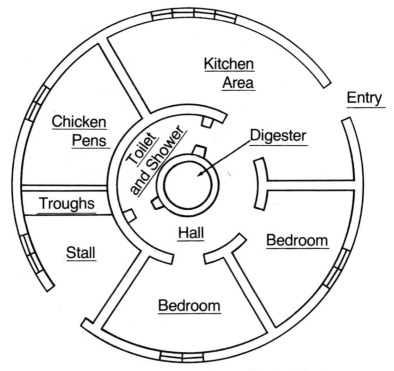

Figure 2. Top view Diagram of a Dwelling Unit for four.

day. In the digester, fermentation once established continues on a steady basis, as does gas production.

Special care will have to be exercised to avoid unnecessary loss of useful components. Therefore, all solids, liquids, and gases must be recycled or consumed. Complex substances are decomposed in the digester. Products of this decomposition are organic acids, ammonia, CO_2, and methane. The methane is stored for use as needed; under slight pressure, it is used for cooking.

The addition of the nutrients to the digester displaces soluble substances into the algal culture, where the latter serve as a base for algal growth. Carbon dioxide, formed by the combustion of the methane, is vented by convection to the algal culture, where a part of it is used as a carbon source by the growing algae.

Algal slurry fed to the cow constitutes its sole source of drinking water, which forces it to consume algal protein in wet form. Algal slurry not consumed by the cow is removed from the trough, and spread over the sand beds. The dewatered and dried algae can be used on the site for chicken feed, also to augment the algal slurry feed for the cow, or sold.

Using the space below the circular algal culture tanks as living quarters gives shelter to both the animals and humans. The algal culture and digester provide a buffer against rapid temperature changes, while the metabolic heat given off by the occupants supplies some warmth to the algal culture and digester during cool periods.

On the basis of past experience, the system can be expected to provide an ample and hygienic environment for a family and its essential livestock. The unit can be constructed of local materials, or perhaps can be prefabricated for import. Because it is largely powered by sunlight energy, the feasibility of such a system is greatest in tropical regions of the world, although it can be of use in other areas during the summer period.

The advantages of the system we have described here are:
1. the creation of a highly livable system for the occupants;
2. the establishment of an efficient and hygienic waste management;
3. the recovery of valuable nutrients from wastes.

A preliminary economic analysis indicates that a gross income of between $250 and $1,000 a year could be realized by operating the system. Operation costs should range from $50 to $100 a year. If only the lower income level ($250) were attained, a substantial subsidy would probably be required. On the other hand, if the higher ($1,000) income level is achieved, the unit probably would be economically attractive.

New Process Increases Coal Efficiency

•

Glenn Kranzley

With nuclear power plant construction lagging and the government-research-industrial establishment giving only nodding attention to the development of solar and wind power, the biggest immediate hope for developing new energy sources is still fossil fuels.

In 1975, $185 million will have been spent on various projects in coal gasification and liquefaction, but the earliest hope for completing any of these is the early 1980s.

Meanwhile, the Ilok Powder Co., a research cooperative in Washington, D.C., claims it has the technology right now to produce a coal fuel which:

—Is more efficient, producing 4.5 barrels of fuel from a ton of coal, compared with a yield of 1.5 barrels in gasification.

—Will burn without pollution, the sulphur, ash, pyrites, and all inorganic compounds having been removed during processing.

—Can be combined with fuel oil or gasoline for a colloidal fuel which could be substituted in diesel engines or for a variety of industrial uses.

—Also has uses as activated carbon for treating polluted water, or as carbon body filters which could be ingested and would filter carcinogens from both humans and animals such as cattle.

The secret of Ilok's grinding process which reduces coal to four micron particles. (The human eye can see no smaller than 40 microns.) Because of the size, the particle burns completely, leaving no residue. Since nothing but carbon is left in the particle, nothing pollutes, Ilok claims.

Glenn Kranzley is a reporter for the Bethlehem Globe-Times, of Bethlehem, Pennsylvania.

The process was invented in Germany in 1936 by the late Ing. Hans Rohrbach and Dr. V. Stephen Krajcovic-Ilok. He emmigrated to the United States in the 1960s, bringing the patents with him. He now is president of Ilok Powder Co.

If Ilok fuel is so good, why isn't it being used? The reasons are complex, and involve the way in which business is done when millions of dollars are involved.

William Talbert of York, Pennsylvania, an Ilok consultant, has been visiting industrial researchers trying to get them to back construction of an Ilok plant at an estimated cost of $100 million for building and first year operation. He's been seeking to interest those businesses which consume great amounts of fuel and therefore would benefit most, such as Bethlehem Steel Corp., Caterpillar Co., Pennsylvania Power & Light Co., and Mack Trucks, Inc.

So far, however, the only contract Ilok has is with the Georgia Power Co., which will help fund the plant, and then will buy 85,000 tons of processed coal per year.

Talbert said the problem is that before an industry will finance such a project, they insist on full technological disclosure and some sort of control of the project. All Ilok is willing to offer at this time is that backers will be first in line to buy the magical coal when it is produced.

No agreements have been made with ERDA for government funds for two reasons: first, ERDA's mission is for research, not building industrial plants; and second, Ilok's lawyers feel they may be unable to protect the secret patents while doing business with the federal government.

Talbert is discouraged but not hopeless. "It's a matter of time. As the cost of oil goes up, our process looks better and better. And when there's nothing left but high sulphur coal, we'll be the only ones who can process it," he says.

Geothermal Energy: A Power Crisis Remedy?

Interest is growing in many quarters over the possibilities of geothermal energy—generating electricity by tapping natural sources of underground steam—as an answer to the energy crisis.

In Congress on June 28, 1972, Senator Mike Gravel (D-Alaska) lamented that only $2.5 million had been budgeted for geothermal energy in fiscal 1973.

"This is absurd," he said, "since our geothermal hot-water resources might offer a quick way to produce a lot of electricity. Geothermal hot-water plants can be built in a year or two, and they are safe."

Up to now, installations of this type have been restricted to locations where underground reservoirs of water heated by molten rock exist naturally, and research is continuing to try to improve the capacity of the process, bringing it into line with economic needs. Two engineers in York, Pennsylvania—J. Hilbert Anderson and his son—are convinced that "they have already solved both the economic and the environmental challenge of this technology," according to Gravel.

"For instance," he said, "their capital-cost estimates on a vapor-cycle plant are lower than other estimates, their kilowatts-per-pound-water-used are better, their new air-cooled condensers require no cooling water, and the underground water is returned underground in a closed cycle."

But scientists at the Los Alamos laboratories in New Mexico feel they have found a way to expand the potential of the process.

By drilling into hot, dry underground regions, pumping cold water in and letting the heat of the inner earth take over, they say, sufficient boiling water and resulting steam should be produced to power electric-producing turbines. If they're right, geothermal energy plants could be placed anywhere that hot, dry rock is available within 25,000 feet of the surface of the earth.

In a recent request to the federal government for further research funding, the Los Alamos scientists concluded that "the development of shallow, dry geothermal reservoirs appears to offer the real possibility of very large amounts of clean, cheap power for a very long time." They

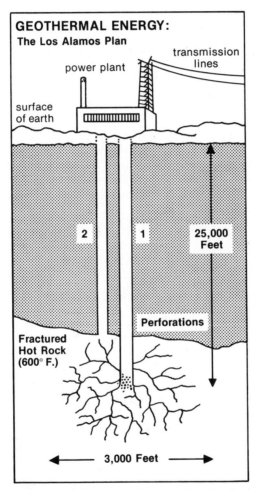

GEOTHERMAL ENERGY:
The Los Alamos Plan

Geothermal plant: Los Alamos scientists feel that by pumping water into the earth (1), pressure would crack rocks below, water would be heated and then would rise to the surface in a second pipe (2), in the form of steam, to power electricity-generating turbines.

further estimated that the continental United States contains enough dry, geothermal reserves "to satisfy this country's electrical energy requirements for several thousands of years."

In a recent interview with the *New York Times*, Dr. Morton C. Smith, Los Alamos research director, said drilling operations could be performed with current oil field-type rigs, and proposed a test drilling of about 15,000 feet into the rim of Jemez Caldera, an extinct volcano near the laboratory. The estimated cost of the test would be $11 million over seven years.

The water, according to the scientists, would have to be heated to about 600 degrees F. to operate a power generation system. Other techniques being studied might be able to bring that figure down to 340 degrees F.

To establish a geothermal energy production plant, a hole would be drilled to the appropriate level and a steel pipe inserted, perforated several hundred feet above the bottom. A high-pressure pump would force water into the well at pressure of about 7,000 pounds per square inch through the perforations, and crack the surrounding rock.

A second hole would be drilled about 20 to 30 feet from the first, also connecting with the cracks formed by the initial hole. The water pumped through the first hole would circulate through the cracks, be heated, and then be forced through the second hole to a generating plant at the surface.

The difference between the temperature of the water being inserted and that being withdrawn would be sufficient to maintain circulation by natural convection, the scientists say. Additional pumping would be used only when necessary.

The heat being removed from the rocks will cause them to shrink, which is expected to cause new cracks and expand the capacity of the installation.

But, as Barry Commoner has said, in environment there is no such thing as a free lunch, and possible problems are being investigated. Research is underway into whether the cracking of the underground rock could touch off earthquake activity, or whether depleted wells could cause the collapse of the surface. Soluble minerals withdrawn with the heated water also may become a problem, corroding the system.

If no serious drawbacks are discovered, however, the system could be in operation within 10 to 15 years. And the United States could have a good, clean source of electrical energy.

The scientists' expectations of success are based on computerized mathematical models of the project.

The cost of energy produced in this manner has been estimated at somewhat less than that from a modern coal, oil, or nuclear installation.

Overhead—maintenance and labor—would be significantly lower, according to the Los Alamos group.

Editor's Note: The National Science Foundation estimates that the United States could generate 137,000 megawatts of electrical power using geothermal sources by 1985. The proposed amount of funds allocated for geothermal research for 1976 by the U.S. Government is $91 million, an increase almost 40 times that of 1973. Yet, federal funding for this energy source lags far behind other energy projects such as nuclear research, which has $1.4 billion earmarked for its development.

Electricity from Composting Leaves

The only clues that one particular compost heap is different from the others made by Walter Lentz of Greece, New York, are two electrical wires. The wires go from metal conductors buried in a yard-square section of the pile to a foot-square plexiglass box inside his garden nursery office which houses a motor. The sign there reads: "This motor is running on electricity produced from compost." Explains Mr. Lentz, who turned his horticulture hobby into a business when he retired after 35 years as a Bausch & Lomb researcher: "The motor's been running day and night for four weeks. It's only producing about four to six volts, enough to recharge a battery. But with a series of storage batteries, I could store enough volts to power my lawn mower and other motors, and with a series of composts I could increase the voltage and store enough energy for bigger motors, even lighting the house and office."

Discovery of Monopole:
How Long Before Magnetic Energy

"The country that rules magnetism rules the universe."
—Dick Tracy comics

The discovery of the basic unit of magnetism, announced in August 1975, by a team of particle physicists at Berkeley, will—if the discovery is real—turn out to be one of the great scientific finds of the century, akin to Einstein's discovery of relativity or Max Planck's outworking of the quantum.

According to Walter Sullivan's article in the *New York Times*, the monopole—which is the name given to the particle—is 137 times heavier than an electron, a number predicted by prescient physicists in the early 1930s.

The implications of the discovery are immense, to understate the case. While an electron is a unit of electricity which creates a magnetic field when it flows, the monopole is a unit of magnetism that creates a field of electricity as it flows. Monopoles can be used, among other things, to generate more monopoles. A flow of monopoles could generate enormous quantities of electricity by tapping directly into the inexhaustible power of the earth's magnetism.

What is needed now is an apparatus to produce and isolate monopoles so they can be used. One physicist, commenting on the discovery, said that a monopole engine in a ship would enable the ship to be pulled across the oceans by the earth's magnetism. As we said, the implications are immense.

Alcohol from Cellulose—
An Energy Breakthrough

•

David Gordon Wilson

The immediate purpose of my testimony is to set the exciting developments of Dr. Mary Mandels and her fellow scientists and engineers at
the U.S. Army Natick Laboratories, of improved processes for the production of fuels and foods from solid wastes, into an engineering and
policy perspective.

Firstly, with regard to the potential new technology which the Natick
Laboratories' developments have made possible, the basic discovery
was of an enzyme which could break down virtually all forms of cellulose. A mutation of this enzyme was then produced having greatly
increased activity and productivity.

This new enzyme makes a long-used process for the conversion of
cellulose into glucose much more attractive. Glucose can be used as a
base for the production of certain foods, animal feeds, and fuels, particularly alcohols. Alcohol is particularly interesting because it has
wide uses in industry, and it can also be added to gasoline in amounts
up to 10 percent without changes in engine design or carburetor ad-
justment being necessary. Currently available sources of cellulosic
wastes would, if converted to ethyl alcohol, provide close to 10 percent of current gasoline consumption in the United States, so that the
match of potential supply and potential demand is good.

There are, of course, many other ways of extracting the energy in
organic wastes. The wastes may be burned in a steam-raising incinerator, and the steam may be used to heat buildings, power turbines,

*This article is testimony given by David Gordon Wilson, professor of mechani-
cal engineering at MIT, to the Congressional Joint Economic Committee's Sub-
committee on Priorities and Economy in Government, on May 20, 1974.*

or drive air conditioners, as is being done in most of the modern European and U.S. incinerators. The wastes may be burned in a gas-turbine cycle, for instance that under development with EPA funding in California. Solid wastes may be classified to drop incombustible fractions out, milled, and burned with coal in a regular electric-utility boiler, as is happening in St. Louis, also with EPA funding assistance. The wastes can be milled, classified and bricketted, and sold as a sulphur-free solid fuel, which is a current commercial development. Municipal, agricultural, and livestock wastes can be composted and one use recently investigated has been as 'compost-fuel'—this is a commercial development which has been recently tested by the Department of Agriculture. Solid wastes may be anaerobically decomposed to produce fuel gas—both EPA and the Bureau of Mines are supporting work in this area—and the same agencies as well as private industry are supporting or developing a large number of pyrolysis processes, which produce solid, liquid, and gaseous fuels and other by-products from solid wastes.

And cellulosic solid wastes can be converted into secondary materials, perhaps with a greater energy saving in some cases than if they were used in their place. Old newspapers can be made into new newspapers, into tissue, or containerboard, or building board, or roofing materials, and so forth.

The vital question is, then, which of these many processes for the reuse of solid wastes is the best, and what value system shall be used to judge the best? Should the government choose one or more processes and help them along by subsidizing research or development?

Current State of Non-Polluting Energy Technology Research

•

Stewart Herman

Few people still expect that technology will guarantee a future of material ease for Americans, let alone the rest of the world. But technology is continuing to develop primarily through corporate and governmental research, and if this research were focused more strongly in the search for environmentally sound energy alternatives, it could benefit society greatly.

Among emerging energy technologies such as coal gasification and the breeder reactor, some of which have aroused considerable concern over their environmental effects, are half-a-dozen technologies known collectively as "direct energy conversion," where the conventional heat cycle of energy conversion is avoided, and waste and thermal pollution consequently greatly reduced.

Perhaps the best opportunity for surveying who is involved and what are the recent advances in direct energy conversion was the fifth annual Direct Energy Conversion short course organized by Dr. Charles Backus at Arizona State University. In a weeklong series of lectures during the second week of January 1974, the managers of large corporate projects discussed not only how the technologies work, but potential pollution problems and prospects for widespread use of the technologies in the future.

Conventional energy conversion to produce electric power requires

Stewart W. Herman is a researcher for INFORM, a nonprofit public interest research group which is analyzing the major corporate projects in alternate sources of energy in order to report on the priorities of U.S. industry in developing environmentally sound new sources of energy for the future.

heat to make steam to drive a turbine which then runs a generator. Less than 40 percent of the heat produced by burning coal, other hydrocarbon fuel, or by a nuclear reaction, is converted to electricity. The rest is discharged to the environment as pollution in the form of excess heat.

Direct energy conversion avoids the conventional steam turbine and generator cycle by employing relatively recent technologies to break electrons off their orbits around atomic nuclei and "collect" them directly into an electric current. Among direct energy conversion techniques are such familiar names as solar cells and fuel cells, which turn solar and chemical energy, respectively, directly into electricity, and such less familiar ones as magnetohydrodynamics (MHD), electrogasdynamics (EGD), thermoelectrics, thermionics, and direct conversion from fusion reactions.

Research and development of these technologies is diffused throughout the country in dozens of large and small programs sponsored by government, universities, and private corporations. The short course in Arizona focused on six major DEC technologies, with corporate managers from the most advanced projects reporting on each one. Following is a short summary of the status of each technology:

Solar Cells: On a flat opaque surface of silicon, cadmium sulfide, or other special material exposed to light, electrons are knocked off the "latticework" of atoms by photons—units of light energy—and collected at various points to provide an electric current.

Prospects for widespread future use are uncertain because of very high manufacturing costs. Silicon cells now cost $20 per watt of output, or roughly 60 times as much as the construction cost of a conventional power plant per watt. Cadmium sulfide cells (often used in camera light meters) could be made for $5 per watt, according to Paul Rappaport, director of materials research at RCA Labs.

While manufacturing costs are high, operation of the solar cell requires no external heat source and produces no pollution.

Fuel Cells: Similar to a battery, but continually recharged by entering fuel and air, the fuel cell breaks down natural gas or some other hydrocarbon into carbon monoxide and hydrogen, which combine chemically with oxygen to form water while releasing electrons. The fuel cell produces no pollution—the only by-products are water vapor and carbon dioxide. However, the fuel cell requires fossil fuel, and consequently will abet the depletion of natural resources, although at a far slower rate than conventional steam turbines. An efficiency of 55 percent is claimed for the fuel cell by Dr. William Lueckel, director of the Pratt and Whitney Aircraft Advanced Projects, compared to the 40 percent efficiency for conventional power plants.

Pratt and Whitney of Connecticut has brought the fuel cell almost to the point of commercial use in programs sponsored by nine electric and

28 electric and gas utilities. Commercial use in decentralized power plants is scheduled for 1978.

MHD: In an MHD generator, coal, oil, or gas burns in a rocket engine-like chamber, forcing superheated, ionized "exhaust" gases down a narrow channel at supersonic speed. A strong magnetic field envelops the channel, and electrons in the hot gases are collected by electrodes in the channel walls.

Fuels are efficiently burned in an MHD generator, but the high temperatures involved encourage the formation of nitrogen oxides, creating a pollution problem while eliminating sulphur oxides.

Richard J. Rosa, director of the MHD Generation Project at Avco-Everett Research Labs in Massachusetts, the major center of MHD work in the United States, noted that an experimental MHD device last year generated electricity at a rate of 250–325 kilowatts for 85 hours.

Thermoelectrics: A semiconductor is a material which loses resistance to the flow of electrons through it as it is heated. Heating one end of a semiconductor causes electrons to flow towards the other much cooler end. Completion of the circuit back to the hot end allows an electric current to flow. Such units, known as "thermoelectric modules," have been developing rapidly from use as controls on gas ranges, to space power plants, and most recently to specialized commercial uses in remote areas at low power levels.

According to Dr. E. Hampl, director of thermoelectric materials research at the 3M Company, recent breakthroughs in semiconductor research have made possible compact generators in the kilowatt range using materials which can operate more stably at higher temperatures, hence with greater efficiency.

Thermionics: A thermionics generator operating in a manner similar to a radio or TV tube also uses heat, but to "boil" electrons off a flat plate. The electrons are collected on a much cooler plate only a fraction of an inch away, and complete an electric circuit back to the hotter plate. Efficiency is low, and the level of research funding is even lower. Gulf General Atomic shut down its thermionic program last year— outside of university-sponsored work, only Thermoelectron Corporation of Massachusetts and Dr. Ned Rasor Associates in Ohio (both sponsored by the National Science Foundation) are continuing research.

Since there is excess heat in conventional power plant boilers, the concept has developed of "topping" conventional systems to tap this "free" heat. MHD, thermoelectrics, and thermionics are all candidates. MHD, according to Richard Rosa, could boost overall steam power plant efficiencies to 50 percent or more. A system using conventional gas and steam turbines in combination being developed by General Electric claims similar or greater efficiencies, however.

Power from the People

•

Robert Rodale

There exists in modern societies an alternative source of energy that has been almost totally overlooked—if not actually eliminated in a misguided effort to improve the human condition. I am referring to muscle power as the energizing force to perform a wide variety of tasks.

Vast amounts of potentially useful power are going to waste in the muscles of Americans and people of other highly industrialized countries. A quick look at the nutritional imbalance of our average diet reveals a clue to the extent of this power. On the average, Americans eat 3,000 calories a day—calories being the most commonly understood index of food energy. In fact, our diet is loaded with foods that are far richer than necessary in food energy—sugar and refined carbohydrates in particular.

Not enough of those calories are taken out of our systems through exercise. Again using averages, a typical adult doing desk work and riding almost everywhere in cars burns up only 2,000 calories a day. What happens to the difference? It is not exactly wasted. It most certainly does not disappear. Worse, it is converted into fat which people are forced to carry around in ever-accumulating amounts as they get older. Ultimately, weight accumulates to the point where the extra work of carrying it around consumes the unnecessary calories that are eaten each day. The result is a balance of obesity—a life sentence to fatness and ill-health.

At this point I am tempted to make a further simple calculation to try to compare the total of unused human energy with our consumption of fossil fuel. But I will resist that urge. Obviously, we are burning coal, oil, and gas at a rate that dwarfs the Btu's in all the doughnuts, cheese, sandwiches, and other foods that we eat. That is not entirely the reason I resist making the comparison, though. Quite simply, the notion that

51

we need large amounts of power to live happy and successful lives is false. There is a limit to the contentment and satisfaction that use of mega-quantities of power can buy, and we passed that limit some time ago. As both a nation and as individuals, we are now letting the momentum of our past real needs for additional productive energy propel us into ranges of power use that have the ability to do real harm to the fabric of our lives. Our society, in fact, could well be burning itself out.

We can live happily while using much less energy than we are presently consuming. And we can also do considerable useful work with our muscles, while benefiting ourselves in many ways. I am not saying that a future happy society can be built in America on a foundation of muscle power alone, but we can use muscle energy to at least help satisfy our power needs. Certainly we should not overlook any underused energy source, especially one that is non-polluting and even has the potential to build health.

Once we realize how valuable muscle power is, we can increase its efficiency tremendously by the proper design of equipment. As all of us learned in school, leverage can multiply the ability of a given amount of force to do useful work. Given a long lever and a suitable fulcrum point, one person can lift a tremendous weight. The principle of leverage can also be used to make a variety of hand-powered jobs easier. A spade uses leverage to make digging easier. And leverage is used to make wrenches and cutting tools more effective.

The usefulness of personal power also can be expanded tremendously once we start thinking of human effort as being more than just "hand work." Actually, our hands and arms are valuable more for their dexterity than for strength. Most people use their upper body muscles so little that by early middle age the arms and shoulders have lost much of their strength.

Legs are an entirely different story. The thigh muscle—the quadriceps—is the largest and most powerful muscle in the body. It is the reservoir of tremendous strength, even in sedentary people. Everyone does some standing and walking, and only a little such effort is enough to keep the leg muscles fairly well conditioned. In our legs is truly the power of the people. We can use that power cheaply to do a variety of useful jobs, while improving our health at the same time. Vigorous leg exercise stresses the heart and lungs, improving their efficiency and promoting long-term health.

The bicycle mechanism is the key to most efficient use of leg power. Because a person rests while riding a bicycle, the legs are relieved of the effort of either standing or supporting the body's weight while walking. (A muscle doesn't have to move to use energy and become fatigued. When leg muscles contract to support weight while a person stands, energy is used.)

Bicycles also provide leverage, which magnifies the power of the legs. Both those factors combine to make bicycling roughly four times more efficient than walking as a means to transform food energy into distance covered. And there are still other reasons why bicycling is so efficient and widely used. The smooth rotary motion of pedaling puts very little strain on the joints, and can (with training) be continued for long periods. The usual peddling speed—60 to 80 revolutions per minute—is just the right rate of action to get maximum efficiency from the leg muscles.

How much power can a person produce, using the bicycle mechanism? Studies have shown that a moderate pedaling rate yields somewhat less than one-tenth horsepower (74.6 watts). Of course, the rate of power output will vary, depending on strength and endurance. Some professional cyclists have produced up to one horsepower, and a recent Dartmouth College study showed that the average "utility" cyclist—going about 8 miles per hour— generates one-twentieth of a horsepower, or 37.33 watts.

Three years ago, I was able to visit the mainland of China as a member of a delegation of journalists, and wrote a series of articles in *Organic Gardening and Farming* about the life of the people there. The thing that impressed me most was the way the Chinese people were operating a large, and on the whole, successful society using only tiny amounts of energy. For example, I found that the Chinese at that time were producing everything needed by their 800 million people with an amount of energy equal to that used in the United States only for air-conditioning. In every aspect of their society, I saw extremely conservative and efficient use of energy.

Bicycles are the main mode of personal transportation in China. They are used not only to get around, but to carry goods. I saw heavy loads of all kinds of items being carried on bicycles. Also, trailers attached to bicycles were widely used.

Back in 1973, I was not quite as interested in pedal power as I am now. I didn't try to search out use of pedal-powered stationary machines. Also, the restrictions on travel in the People's Republic of China made looking for items of personal interest quite difficult. The official guides are especially reluctant to take you to see old-fashioned things. They would prefer that you see modern engines rather than "obsolete" pedal-powered machines.

During a recent trip to Taiwan, the old Formosa, I had a chance to compare energy production and use with what I had seen in China. Taiwan uses more power per capita to run its society than does China. Some people here own cars, and many have motorcycles. There are more lights, more appliances, and more mechanization of all kinds. There is air-conditioning in the modern buildings.

Yet bicycles are still an important means of basic transportation here, as they are in all developing countries. Millions of people in Taiwan use them to go to work, for shopping, for all their local travel needs, and to carry goods. There are no pedal taxis here, as are common in India, but many thousands of pedal-powered tricycle "trucks" are in use. I saw some of them carrying unbelievably large loads. The gear ratio on these tricycles is low, so each revolution of the pedals moves it a relatively short distance. That makes pedaling easier. The hard part is getting started. Once momentum is gained, even a loaded tricycle-truck can be moved along fairly easily, thanks to low gearing.

Several times during my trip to Taiwan, I asked agriculture experts here whether pedal-powered, stationary machines are still in use. Foot-powered threshing machines used to be common, but I was told that almost all of them have now been motorized. The old wooden man-powered irrigation pumps have been replaced by engine-driven pumps. But I didn't strike out completely in my search for Chinese pedal technology.

One of the most productive visits in Taiwan was to the Asian Vegetable Research and Development Center, located near the South Taiwan town of Shanhua. While I was there, Merle R. Menegay, the agricultural economist of AVRDC, helped me find a working example of pedal power currently in use in Taiwan.

"Are any foot-operated machines still in use here?" I asked Merle over lunch the first day of my visit.

He thought a few minutes, then said yes, he had seen the old type of all-wood, foot-driven pumps being used to raise water from one level to another at a salt works over by the coast. The next morning, Charles Taylor, the Center's public information officer, drove with me to the little town Merle Menegay had mentioned. We found the place easily and took several pictures of the pumps. They weren't being used when we were there, but we looked them over carefully and saw that they worked well.

Those foot pumps were made in the ancient Chinese way, apparently without the use of any metal parts. The water being pumped lubricated the mechanism. Working the pumps was quite easy.

Why were such old pumps still in use in a rapidly developing country like Taiwan? Looking over the water-moving requirements of the salt flats, I saw that those wood pumps were a perfect application of appropriate technology. They were the best way to get the job done, for the least cost. The water had to be raised only a couple of feet, and the depth of water in each of the evaporation ponds was only a few inches. Probably an hour or two hours' work every few days would put enough new salt water into each of the ponds. Using gasoline pumps would be more costly, and might also pollute the salt with oil residue.

After leaving Taiwan, I stopped in the Philippines for a few days to see the International Rice Research Institute at Los Banos, Laguna. One of their big problems is to make small farms more efficient, through the use of better machines as well as improved plants. Getting water into the rice paddies more easily is a problem they are now working on. The average paddy is less than an acre in size, and motorized pumps are often far too expensive. Very often, water is carried from the ditches up into the paddies using buckets and a bamboo carrying pole.

Engineers at the International Rice Research Institute (called IRRI for short) have developed a foot-powered pump that makes that method obsolete. It is a small, lightweight and inexpensive device that will lift large quantities of water several feet, using only moderate amounts of "people power." The operator simply stands on two footrests at either end of the pump and rocks back and forth. That effort compresses a diaphragm, which forces water from the outlet valve. By operating the pump in a rhythmic manner, a continuous flow of water is pumped.

Bart Duff, an agricultural economist at IRRI, estimated that the foot pump will cost only $25 to $30 to make. IRRI gives designs for its machines free to fabricators throughout Asia, who make and sell pumps, tillers, threshers and other implements to local farmers. The original foot pump design used a rubber bellows, but that wore out too soon. Now a diaphragm made from a section of truck inner tube is the heart of the pump, and tests show that it lasts a long time.

Even more exciting than the pump were the foot-powered trolleys that are the basic means of getting to work for many IRRI staff people. They are simple devices yet extremely efficient.

The frame of the trolley is made of wood, with bamboo slats used for the floor. There is a bench for the passengers, a small platform at the rear for the operator to stand on, and also a bar at the back which the trolley-pusher holds onto to keep his balance. A brake, which rubs against the train rail, is also built into the trolley. Wheels are made from roller bearings salvaged from junked Mack truck wheels. An extra set of four bearing-wheels mounted at each corner of the trolley keeps it on the track.

Young men push these trolleys from early morning to late in the evening, carrying people from villages along the track to their jobs at the Institute. They charge a small fee for each passenger, plus an additional charge for packages. Of course when a train comes along, they get off the track fast. Fortunately, the trains on that line don't make much speed, possibly because the track is not in good shape. There is also a well-defined procedure for deciding which trolley gives way when two meet head on. The one that is more heavily loaded—counting both passengers and packages—stays on the track.

I understand that trolleys like these are used all over the Phillipines, wherever a rail line goes where people want to travel. They impressed me as being a very efficient and sensible means of transportation—combining the strong motive power of human leg muscles with the low-friction rail roadway.

Even though my muscles weren't conditioned for that kind of work, I was able to pedal the trolley quite easily, carrying both one and two passengers. And there was a slight headwind, too. One push with my leg would move the trolley quite a few yards.

I don't recommend that we use rail trolleys widely here, except on abandoned rail lines. The idea of having a speeding train catch up with you from behind while you're tooling along on a trolley isn't too appealing. But there's also no escaping the fact that the Philippine rail trolley—like many other pedal-powered devices—has the energy efficiency we need badly today. It's just one more example of the effectiveness of power from the people.

Part II

•

Agriculture: Energy Asset or Liability?

The "Green Revolution" was touted in the late 1960s, when the program was started, as the means for saving the world from starvation by producing bumper crops with the use of abundant, cheap petrochemicals to make fertilizers, pesticides, and herbicides and to power the huge machinery necessary to spread these chemicals. At the same time, organic farming was scoffed at as being not productive enough to feed the earth's billions. Secretary of Agriculture Earl Butz summed up the feeling at the time when he said, "If some environmentalists had their way, 145 million Americans would be eating organic foods, and 60 million would be starving."

Many of the arguments for this kind of highly mechanized agriculture are vulnerable, and it took the first signs of energy shortages to point this out. Production per man-hour had increased with mechanized agriculture, but production in terms of energy expended had been reduced radically. Agriculture was (and still is) an energy sink. When we add up the quantities of petrochemicals needed for fertilizers, herbicides, and pesticides; fuel for farm equipment and transportation to market; and energy to process and package the food, we find that it takes approximately five calories of energy to put one calorie of food on a dinner table. In contrast, "primitive" farming methods, like those practiced in parts of the Orient, according to some studies that have been done by Marvin Harris, of Columbia University's Anthropology Department, are able to produce 50 calories of food energy for every calorie used. The energy expended here is human energy, about the only kind of energy they use, since most of the work is hand labor and their fertilizer is primarily night soil.

In 1975, Dr. Barry Commoner and the Center for the Biology of Natural Systems at Washington University in St. Louis, began a comparative

57

study of organic farming and conventional farming which is quite dependent upon chemicals. Although the study has not been completed, the early available data is quite interesting. While both types of farms had approximately the same crop yield per acre, those relying heavily on chemicals have been found to use $16 more per acre for fertilizers than organic farms.

Agriculture as it exists now in this country is our biggest energy consumer—and perhaps the most wasteful energy consumer. And yet, as we attempt to show in the following articles, it could be an extraordinarily valuable net energy producer.

Low-Energy Agriculture

•

Jerome Goldstein

The 700-acre organic farm of Iowan Ralph Englekens is yielding 185 bushels of corn per acre on his best land and 125 bushels on his poorest land. These figures may come as quite a shock to many agricultural experts who stoutly proclaim that the present levels of chemical fertilizer applications are absolutely essential to maintain yields.

The average conventional corn grower in Iowa uses 110 pounds of nitrogen, 60 pounds of phosphorus, and 56 pounds of potassium in chemical form, on each acre. The total bill for agricultural chemicals in Iowa in 1969, including fertilizers, herbicides and pesticides, came to $203,097,414—in 1972 these costs were up to about $30 an acre.

By contrast, organic farmer Ralph Englekens spent $3.25 per acre for fertilizer in 1971, the same amount in 1972, mostly for rock powders and soil amendments. The bulk of his fertilizers, put down at the rate of 15 tons per acre, is manure from his many farm animals.

Englekens is by no means alone among farmers who have developed other ways of bringing nutrients to their crops than simply using such material as anhydrous ammonia, superphosphate, et al. The simple fact is that fertilizer and pesticide bills have gotten too high for farmers not to be looking at cheaper alternatives.

Aside from the expense, the energy crunch is hitting directly at the manufacture of agriculture chemicals. Serious shortages are predicted for next spring. Huge amounts of electricity are needed to make commercial fertilizers. Nitrogen fertilizers are made from natural gas, which is in critical supply.

Jerome Goldstein is the Executive Editor of Organic Gardening *and* Farming *and is the Editor of* Environment Action Bulletin *and* Compost Science.

The farmers who are leading the trend away from strict reliance on energy-draining commercial fertilizers are the organic farmers. But there are problems when you turn to bulky organic materials to put nutrients into your soil, instead of opening up some bags of 10-10-10.

"Organic farming is more than just spreading manure on the land," explains Mike Scully of Buffalo, Illinois, who stopped using chemicals on his own 400 acres and the 300 acres he rents. "You have to get an ecological balance." And that balance begins with the soil and continues through the crops and animals grown on that farm.

Farmers who use sludge are faced with similar needs. In his report, "Soils as Sludge Assimilators," Jim Evans of the USDA recalls how in 1967 he skeptically "observed sludge being spread and saw previously sludge-treated lands representing a range of soil types and topographic and land-use conditions from the Allegheny Mountains of north central Pennsylvania to the highly productive farmlands of the lower Susquehanna Valley in southeastern Pennsylvania. I found it difficult not to be convinced by what my eyes saw, my nose did not smell, and my ears heard from the enthusiastic farmers who were spreading sludge on their pastures and croplands. My impressions were summed up in the article 'They Spread Black Gold on Their Fields' which appeared in the February 10, 1968, issue of the *Pennsylvania Farmer*. The expression, 'I call it black gold' was used by one farmer in Elk County when asked to give his opinion of the digested sludge he was receiving free of charge from the St. Mary's sewage plant."

But even "black gold" has to be analyzed, as do the results of applying such materials to the soil.

The answers needed by farmers who apply such materials as composted refuse, sludge, or manure go beyond the conventional soil tests of the past. One survey taken a few years ago showed that state and private soil testing laboratories in this country tested some four million soil samples. Unfortunately, soil tests have been used largely to sell fertilizer—resulting in overfertilizing and heavy expenses to farmer and the environment.

To satisfy the needs of organic farmers and others, independent laboratories are offering their services to analyze soil, feed, water, and even livestock blood samples. For example, Brookside Farm Laboratory Association in New Knoxville, Ohio, saves its farmer clients money by reducing fertilizer use without reducing yields. For as little as $2 an acre, a farmer can get the insurance of sound advice that used to come at 10 to 15 times the cost from purchases of agricultural chemicals.

The emphasis on rotations, applying organic wastes, building soil humus content, etc., is tremendously significant to farmer and non-farmer alike.

Writing in *Science* (November 2, 1973), members of Cornell University's College of Agriculture and Life Science document the critical need for changing present agricultural practices in a report entitled, "Food Production and the Energy Crisis." It's must reading for everyone responsible for waste treatment in this country, whether that waste be urban, industrial, or agricultural.

"Farming uses more petroleum than any other single industry," says the Cornell research team headed by David Pimentel and L. E. Hurd. "Energy is used in mechanized agricultural production for machinery, transport, irrigation, fertilization, pesticides, and other management tools. Fossil fuel inputs have, in fact, become so integral and indispensable to modern agriculture that the anticipated energy crisis will have a significant impact upon food production in all parts of the world which have adopted or are adopting the Western system.

"*As agriculturists, we feel that a careful analysis is needed to measure energy inputs in U.S. and green revolution style crop production techniques.*"

Predict the authors: "When energy resources become expensive, significant changes in agriculture will take place."

To investigate the relationship of energy inputs to crop production, the Cornell researchers selected corn—the most important grain crop grown in the United States. The use of fertilizer in corn production has increased steadily since World War II—reaching in 1970 as a national average 112 pounds of nitrogen per acre, 31 pounds of phosphorus per acre, and 60 pounds of potassium per acre. Pesticide use in corn has likewise risen rapidly, with about 41 percent of all herbicides and 17 percent of all insecticides used in agriculture applied to corn.

While exploring alternatives to reduce energy inputs in agricultural food production, Pimentel, Hurd, and colleagues analyzed the potential in a single organic waste material—manure. And the potential savings are astounding! If just 20 percent of the manure now produced in feedlots and confinement were available for use in corn production, 17 million acres of corn could be fertilized at an average manure application rate of 10 tons per acre. "In addition to saving valuable fuel energy, applying this manure to cropland would effectively recycle these animal wastes," point out the researchers.

In fact, according to the authors, "if manure were substituted for chemical fertilizer, the savings in energy would be a substantial 1.1 million kcal per acre."

In their conclusion, Pimentel, Hurd and colleagues make the following points:

"To reduce energy inputs, green revolution and U.S. agriculture might employ such alternatives as rotations and green manures to reduce the high energy demand of chemical fertilizers and pesticides. . . . By employing combinations of several of these alternatives, we estimate that it would be possible to reduce energy inputs by about a half and still maintain present yields."

The energy crisis has been well publicized. Whatever the causes—political, economic, environmental, people, and so forth—there seems little doubt that energy resources will be *more expensive than ever before*. According to the Cornell researchers, the result should be "significant changes in agriculture."

At some point, those changes on the farm must relate to changes in the city. One way to make sure is to set up a waste connection now. The connection would look something like this:

Cities gather wastes; cities transport wastes to rural locations; resource recovery centers in rural areas compost wastes; composted wastes are made available to farmers.

The advantages of a waste connection are listed below:

1. Cities have a solution to their waste problem—more economical and environmentally sound than any other available;

2. Rural development benefits exist from job potential in waste treatment-and-composting plants (includes transportation and resource recovery effort for other materials not used in agriculture);

3. Farmer benefit in lower fertilizing costs and higher profit potential;

4. Energy savings.

Of course, the complexities involved in a sound and sensible waste connection are enormous. Cities must monitor wastes if the end product (wastes leaving its collection plant) are to be controlled and beneficial for use on farmland. If not controlled, then we'll all continue to go round in circles over whether or not such wastes as sludge are fertilizers that are "threats to our food supply" or "black gold even better than manure."

Perhaps just as critical as monitoring is the need to prepare the waste connection so that the political problems are overcome. For the connection to be made, wastes will have to cross all kinds of boundaries—city, town, county, as well as state. All kinds of government agencies potentially stand in the way—even though all citizens stand to benefit.

And finally—who will *benefit* from the waste connection? The forecast is that the solid waste disposal and recovery market will *triple* by 1980. Is that tripling only going to benefit the same consultants and manufacturers who were pushing incineration and dumping equipment just a few years ago?

Why can't some of those predicted profits help to put some rural areas back into business—or help bring some extra income to those who still want to farm the land? By aggressively developing the right kind of waste connection now, cities would wind up paying less . . . and some of their expenses would help solve problems in rural-and-farm areas as well.

If we don't carefully spell out the type solution we want, then we'll wind up with the kind of "solution" that is far from the best.

Innovation is desperately needed right now to develop the structure for achieving waste treatment solutions that mean low costs to city taxpayers and low energy agriculture to the farmers who grow our food.

One way to help that innovation along is for government agencies to openly declare their desire for such solutions! For example, a state director of agriculture—interested in a wastes-back-to-the-land policy—could set up a special committee whose purpose is to create the framework for an urban-agricultural waste connection. Such a committee should be committed to relating urban wastes to benefit farmers, soils and rural development.

The problems may not get any easier when looked at by a waste connection committee, but at least the members of such a committee will have a picture of what they'd like to see happen when a solution is found. And that just might be a key to problem-solving.

Organic Energy Farms: Will They Free Us?

•

Jim and Peggy Duke

What if, instead of making auto fuel from the fossilized remains of plants (oil and coal), we made it from fresh plant tissue? High-quality, low-pollution fuel is as available from fresh plants as gasoline is from oil.

We contend that by growing plants with conversion to fuel in mind, America could become self-sufficient in energy, utilize some of its organic wastes to build soils, and achieve many other benefits.

Assume we have 62.5 million acres of unused and marginal land available for energy farming. Now let us develop a concept we inherited from the American Indians—the intercropping of a legume like alfalfa with a cereal like corn, adding some organic matter like sewage sludge. Alfalfa, like most legumes, takes nitrogen from the atmosphere and puts it in the soil—nearly 200 pounds per acre. It would take nearly a barrel of oil to manufacture that much inorganic nitrogen fertilizer.

Alfalfa grows well in the cool months, producing enough vegetation to yield the energy equivalent of two to seven barrels of oil per acre. Basing estimates on average alfalfa hay yields, participants at the Fourth Annual Alfalfa Symposium concluded that we could get nearly a ton of leaf protein per acre from alfalfa. This would mean 55 million tons of protein from 62.5 million acres—about 10 times what Americans need in their diet. Residues remaining after protein extraction would yield the equivalent of 250 million barrels of oil in residues. This alone could cut gasoline imports significantly. If we fertilized with sewage sludge, our 62.5 million acres of alfalfa could probably meet

---

Jim and Peggy Duke operate the Herbal Vineyard in Fulton, Maryland. He is a professional botanist.

President Ford's target import reduction of one million barrels a day, but also provide protein for the United States with plenty for export.

As for fertilizer, manure from feedlots on the energy farm will complement the spent mash (left after conversion of plant residues to fuels) and the nitrogen fixed by the alfalfa. Sir Joseph Hutchinson says (in *Nature*: 254, 1975): "Greater savings are to be gained by better distribution of animal manure and by the use of leguminous crops to augment the nitrogen supply."

Interplanting corn in stubble is coming into style as an energy and soil conservation device. Corn rotated with alfalfa has given better yields than corn in monoculture. According to data presented by Dr. David Pimentel and his Cornell associates (in *Science*), it takes about two barrels of oil per acre to grow corn (nearly a barrel for nitrogen fertilizer); however, there is the energy equivalent of 20 barrels of oil per acre in the stalks and leaves, and another six in the grains if we use them for energy.

Corn is one of those productive plants known as C-4 plants, which utilize sunshine, water, and carbon dioxide more efficiently than the typical C-3 plants, producing more "barrels of oil" from a given amount of water and carbon dioxide. Other highly productive C-4 plants include these grasses: bluestem-, buffalo-, columbus-, crow-foot-, faragua-, grama-, Job's tears-, Johnson-, kikuyu-, koob-, lemon-, love-, molasses-, pangola-, para-, proso-, rhodes-, sudan-, and timothy-grasses and millets, as well as sugar cane. These are some of the world's most efficient fixers of solar energy. Some also have nitrogen-fixing bacteria associated with their roots, like legumes. As energy consumers, we need to know more about these energy producers.

Organic matter in the soil is important for the best yields. On a hot, still day in a corn field, the green plants often shut down when their "fuel"—carbon dioxide—is locally all used up. C-4 plants like corn are more efficient at using the available carbon dioxide, but when the last bit is used, no more photosynthesis occurs. Humus or sludge in the soil of the *organic* energy farm is constantly being "digested" by the healthy soil, however, giving off both gaseous carbon dioxide and water, thereby permitting more sugar (or energy) manufacture by the plant.

Even without using sewage sludge, we could expect, conservatively, 0.8 barrels of oil in per-acre energy savings from the nitrogen added to the soil by an appropriate legume; 3.2 barrels from legume residues (after extraction of protein); four barrels in corn grain; and 20 barrels from corn residues, totalling 28 barrels per acre. Put another way, we gain more than a billion barrels of oil a year from putting unused and marginal acres into production as organic energy farms. In 1974, we imported less than 2.3 billion barrels. Since sewage sludge can double some crop yields, it conceivably might double the yields of our energy farms, making our unused lands capable of producing more than two billion barrels of oil per year. By doubling our energy productivity, organic matter could double our chances for self-sufficiency.

In 1973, there were nearly 62.5 million acres in hay and 62.5 million in corn. Putting these 125 million acres into protein-energy farms of appropriate combinations of legumes and cereals, we could generate another 3.5 billion barrels of oil, after harvesting for export or local consumption 100 million tons of legume protein and more corn cereal than we have ever harvested before. Sewage sludge could increase this yield, maybe even to seven billion barrels of oil, in addition to surplus cereal and protein. That is twice the oil we imported in 1974, but *this* energy could be grown year after year. *With the help of organic matter (e.g. sewage sludge) we could become self-sufficient in energy, export protein, and export corn, using just the 187.5 million acres of land that was recently in fallow, corn, or hay.* Our balance-of-payment posture could be improved by more than $20 billion.

In becoming self-sufficient via organic energy farms, we could generate employment for the depressed farming, housing, and automotive industries. More hands will be needed to plant, cultivate, harvest, and process energy crops. Small factories will be needed near the energy farms to convert energy crops into renewable fuels like ethanol, methanol, and methane, all of which generate less pollution than gasoline. Detroit can reverse its slump by manufacturing converters needed to run our cars on renewable fuels. By decentralizing the fuel production process, eliminating the transport of fuel halfway around the world, we will also stimulate depressed local economies and conserve energy in fuel transport, to say nothing of removing the oil producers' fingers from our economic neck. By converting to organic renewable fuels, we generate research and jobs for America rather than for OPEC. Price shifts following such a conversion might make it possible to trade a bushel of corn for a barrel of oil.

Current efforts to deal with our energy problems call for massive use of western waters—water that will be needed by farmers to grow crops. The U.S. Army Corps of Engineers has proposed that it be given the power to determine who will get water, and for what. But the organic energy farm will alleviate both water problems. In fact the water removed during processing of crops for energy and protein could be piped back to the fields. In addition, the buildup of humus in dry, western fields would help to hold the scanty rain.

Strip miners can be expected to apply political pressure to harvest America's coal. If granted the right to mine, the strip miners should convert the torn land into energy farms, interconnected by canals dug with their earth-gouging machinery. Stripped coal could then be barged out and sewage sludge barged in to fertilize and rehabilitate the land. Before long, barges would be hauling renewable fuels to urban centers and urban sludge back to energy farms. Residues from the energy crops would be returned to the soil, either directly or after having passed through livestock or poultry raised on the energy farms. We recommend feedlots in association with energy farms to produce meat protein from energy crop residues, and to produce manure for fertilizer or conversion to methane. We need methane to replace natural gas that has become so scarce and expensive to the farmer.

We conclude that 187.5 million organic acres of balanced combinations of legumes and cereals *could* make the United States more than self-sufficient in energy, protein, and cereal. Potential yields of some legumes and non-legumes are shown in the accompanying chart. Different combinations would serve best in different regions.

POTENTIAL ENERGY YIELDS OF
SOME WEEDS AND CROPS

*(Figures are energy equivalent
in barrels of oil per acre*)*

LEGUMES

Alfalfa 2-7
Alyce Clover 3-5
Beggerlice 2-6
Crimson Clover
    3-8
Gorse 6-12
Mothbean 2-6
Kudzu 2-5
Peanut 1-2
Siratro 1-6
Soybean 6-14

NON-LEGUMES

Corn 16-24
Elephant Grass
    9-38
Great Bulrush 2-40
Hilo Grass 20-93
Papyrus 16-46
Reed 38-39
Rice 20-26
Smooth Cordgrass
    21-24
Sugarcane 6-50
Water Hyacinth 10

*Assuming the energy in 1,320 pounds of dry matter is equal to that in one barrel of oil.

Food Policies That Save Energy

•

Jerome Goldstein

If you care about energy conservation, you should care where your food comes from and how it is grown. What sense does it make to transport a chicken some 1,500 miles to a supermarket—burn up all the fuel that it takes for a truck to travel halfway across the nation—when that same chicken could have been grown by a farmer near you? It makes no sense at all, yet that's exactly what is happening today.

We are wasting all kinds of energy in processing and transporting food unnecessarily. Recognizing the waste along with the need to develop regional food policies that encourage farming, many states are beginning to develop Food and Agriculture Policies. Energy-saving is seen as one of the key goals, but there are other reasons why changes are urgently needed.

Consider Massachusetts:

• More than 85 percent of the food consumed by state residents is imported—97 percent of the meat and poultry, 70 percent of the eggs, 80 percent of the milk, and 90 percent of the potatoes.

• Food costs are 10 percent higher in Massachusetts than the national average.

• Only 6,000 farms exist today, where 55,000 existed in 1945.

• About 20,000 acres of Massachusetts farmland are lost annually to developers.

• Only 35,000 citizens are employed in farming and fishing industries, and the number is shrinking.

Pennsylvania has a similar set of problems, again related to the fact that so much of the food consumed by its residents is imported over long distances.

In 1960, there were 106,000 farms in Pennsylvania, utilizing 12,300,000 acres of land. The figures for 1975 show 72,000 farms utilizing 10,000,000 acres. Thus within 15 years, the Keystone State lost 34,000 farms, and 2,000,000 acres of land went out of production.

Explains Pennsylvania Secretary of Agriculture Raymond J. Kerstetter: "One of the major reasons, of course, is that our small farmers have a difficult time competing in today's marketplace, dominated by large agri-business complexes and huge supermarket chains. Small farmers have been unable to deliver the huge volume and variety of fresh fruits and vegetables demanded by mass buyers. As a result, they often have been bypassed in favor of the mammoth production and marketing operations in the Southwest and Pacific Coast regions."

Regional self-sufficiency is a term much used by the citizen concerned with energy conservation, and the policymaker responsible for rational economic development. These comments on self-sufficiency by J. Roger Barber, New York State's Agriculture Commissioner, illustrate such concern: (Mr. Barber made these remarks before the New York Assembly Task Force on Farm and Food Policy, February 12, 1976) "Lately, I have been talking quite a bit about this whole idea of making New York State self-sufficient in its agricultural capability. I don't mean that we in this state should strive to isolate ourselves from the outside markets, but I do think it is high time that we recognize that the objectives of our state food policies are necessarily different than the objectives of the food policy established by our national leaders.

"Self-sufficiency means that New York State has everything to gain by trying, as much as possible, to feed itself from its own land. New Yorkers shouldn't have to spend extra food dollars on unnecessary transportation and marketing of their food. The regional decline in rail freight service and the resulting reliance on the more expensive mode of trucking, together with the cost of warehousing and storage are expensive items in the shopping cart. As a key state in the Northeast region, we all are aware that New York and its neighbors have been sold down the river by national transportation policies and discriminatory freight rates.

". . . Self-sufficiency means rethinking our goals. What we have got to do is maintain the agricultural industry as a viable, thriving industry so that it can provide for the people of this state. Self-sufficiency means that we try to figure logical ways to assist the consumer, so that food can be purchased from producers that are not far away, and so that the marketing system for that food is as simplified as possible. We know that a farmer, who can sell to his neighbors, takes more pride in his product and is able to deliver a fresher, higher quality product."

As is evident from the above, states are becoming increasingly aware of the need to shorten routes from farm fields to consumers. Again and again, a major statistic cited is that energy use for food transportation has more than tripled in the past 30 years.

Obviously, millions of Americans who right now are practicing energy conservation measures in whatever personal ways they can would support an agricultural and food marketing system that featured lower energy use—especially if and when lower energy use also means lower prices and fresher, more nutritious foods. Direct marketing is basic to an energy-saving food network.

Much of the thrust of state agricultural and food policy programs is to encourage a system of direct marketing—roadside stands, farm markets, pick-your-own, etc. "Massachusetts Grown . . . and Fresher!" say bumper stickers along Boston highways. "Grown in New York State" labels are used on packages of apples, potatoes, and other foods, and also appear in supermarket newspaper ads.

According to Don Cunnion, chief of the Pennsylvania Agriculture Department's Marketing Services, organized direct marketing in the state goes back at least to 1710 when market houses in Philadelphia rented stalls to area farmers. As other communities grew up and expanded, more farm markets came into being. This was particularly true of communities with heavy populations of Pennsylvania Germans. Curb-type or open-air markets were most common in the Pennsylvania Dutch areas.

After years of decline, we are now benefiting from a resurgence of the farmer-to-consumer selling. Today, reports Cunnion, there are between 600 and 700 roadside stands in Pennsylvania. "Your guess is as good as mine as to the number of umbrella and card table operations that come and go along the roadside," he notes.

The last few years have seen the development of what is known as pick-your-own direct marketing. At least 100 such operations now flourish, including both vegetables and fruit.

Farm markets in urban centers are springing up throughout the United States, enabling local farmers to bring in crops to consumers on a regular basis. An agricultural economist at Pennsylvania State University, Jim Toothman, estimates annual sales through such farm markets at $35 million. New ones continue to open up—and each one helps to shorten the energy which is used unnecessarily in transporting foods thousands of miles.

Still another significant force in direct marketing is the rapid rise of food cooperatives, and the federation of such co-ops for the purchase of food directly from producers.

The urgency of developing regional food strategies is recognized by policymakers in and out of government. Several years ago, James Nolfi and George Burrill set up the Center for Studies in Food Self-sufficiency in Burlington, Vermont. Their purpose was to investigate specific areas in which the expensive gap between food producers and consumers might be narrowed, and they began a detailed study of food self-sufficiency and energy expenditure in agriculture in Vermont.

The researchers gathered data on energy consumption in Vermont agriculture, food consumption patterns, the history of agricultural production in the state, and initiated a computer-based, crop-potential, mapping project. Its goal is to identify areas suitable for producing various crops, especially grains and others that were widely grown in Vermont in the past.

Like its neighboring states, Vermont also set up a State Food Commission to investigate citizen attitudes toward food supply and regional independence. Comment Nolfi and Burrill: "By conducting public hearings throughout the state, the Commission is providing a superb opportunity for supporters of food co-ops, farmers' markets, nutritional education, and agricultural diversification to make their views known and considered in eventual legislation."

In early 1976, following a series of public meetings held by state agencies, the Massachusetts Secretary of Environmental Affairs and the Department of Food and Agriculture issued a proposed "Policy for Food and Agriculture in Massachusetts" which listed these objectives:

• To preserve our agricultural land;
• To increase production and processing of local products;
• To promote local purchase of Massachusetts-grown produce.

In looking at ways to achieve those objectives, the Policy places a heavy stress on home and community gardening, food cooperatives, farmland preservation plans, establishing definitions and standards for organically grown foods, and encouraging the development of less energy-intensive forms of fertilizers including the greater use of organic waste. With specific reference to the latter, the Policy states: "This Executive Office will coordinate the efforts of the Department of Food and Agriculture, the Division of Water Pollution Control, and the Bureau of Solid Waste Disposal to explore possibilities for the utilization of organic wastes (such as sewage, manure, leaves, garbage) for agricultural purposes to reduce the cost of and our dependence on imported fertilizer, especially that which is petroleum derived."

Another major change brought about by the high cost and scarcity of energy is recognition of crop rotations. Until recently, one-crop farming has been considered to be the epitome of industrialized food production. Monoculture was hailed as the "King of Efficiency." No more. Now top agricultural scientists blame a narrow crop base for much of the serious soil, insect, and weed problems. Thus they predict major changes toward planting legumes, using manures, and in general making wider use of practices associated with organic farming.

In a series of 1976 articles labeled "The Future Revised," the *Wall*

Street Journal noted how a "worsening of the energy crisis might return older farming technology and production arrangements to favor. Farmers might use less chemical fertilizer and pesticide, more animal manure, and natural pest killers . . . resume crop rotation to preserve soil nutrients . . . buy smaller, less energy-consuming gear."

Of great importance is the massive desire to get science and technology to shape a desirable, energy-conserving future. The world-famous anthropologist Margaret Mead has called for an end to "misapplied technology" that removes people from agricultural land to the city where there are no jobs available. "We have impoverished the rural world everywhere, substituting imported material for natural material." She hopes that basic scientific research will lead "to the kind of applications that we would *prefer* to have in the world in contrast to the kind that we have now."

The lack of innovative research leading to more energy-efficient agriculture and regional food self-sufficiency has been vehemently criticized. In a recent report, "Enhancement of Food Production for the United States," the National Academy of Sciences (NAS) upbraided the U.S. Department of Agriculture for its unwillingness to plan research needs relating to energy, the environment, and social factors.

An important area of research called for by the NAS is an "evaluation of alternative technologies for reducing energy requirements in food production and handling." One specific proposal called for a "focus on ways of decreasing dependence upon chemically synthesized nitrogen fertilizer" and to increase reliance on biologically fixed nitrogen by use of manure and intercropping with nitrogen-fixing plants.

When Sylvan Wittwer, chairman of the NAS Board on Agriculture and Renewable Resources, was asked if he was advocating a return to the principles of organic farming, Dr. Wittwer replied: "Obviously it relates to the so-called issue of organic farming, but it is broader than that. The use of legumes is becoming a lost technology. That and other techniques of nitrogen fixation are vastly lacking in our nation, and we need to use all the resources we have."

With the increasing recognition of how soils with high humus need less petroleum-based fertilizer and less fuel to power tractors when cultivating, there is a new effort to use organic wastes in agriculture. "Let's not refer to our animal and agricultural residues as waste," declared soil microbiologist T.M. McCalla of the University of Nebraska at a training seminar for farmers. "Let's call them resources and start using them. . . . We have learned, through biological control, that microbes can be used to prevent certain plants and organisms from growing. This means, through biological control, we may become less dependent on chemical fertilizers and pesticides."

In a follow-up report to his original research, Dr. David Pimentel and associates at Cornell University have published "Energy and Land Constraints in Food Protein Production." (*Science*, November 21, 1975). After many references to greater human consumption of vegetable protein, energy loss in feedlot cattle production, the need to preserve farmland, and people to work the farms, the authors note that crops have physiological limits in their ability to respond to increased amounts of fertilizers. There are limits to technology just like there are limits to growth, and Pimentel quotes Sir Julian Huxley, who argued that science has been "completely unable to cope with the appalling problems" of the developing world today.

Another indictment of how these high-energy, agri-chemical approaches are wrecking farmers in other nations appeared in the British publication, *New Scientist*, under the whimsical title: "Agronomy to the Rescue."

As seen by Dr. Charles Posner, Guatemala has raised its per capita food production by 11 percent since 1960, but the sad fact is that mass starvation is now a possibility there. The paradox is primarily attributed to a handful of native agronomists who were schooled in U.S., European, and Soviet practices. They all argue that large-scale farming is more efficient than the family farm and that specialization means a higher return than integrated farming. "Following this theory," writes Dr. Posner, "the agronomists encouraged turning over the richest agricultural land in the country for grazing, and also encouraged the clearing of jungle for plantations, which turned the area into a virtual desert."

Now a study in Guatemala shows that the small farm is more efficient than the large farm. "Farms of one acre have maize and wheat yields no lower than farms of more than 1,500 acres, yet the energy input on the larger farms is three times as high as the tiny farms," explains Dr. Posner. "Yields for rice, beans, potatoes, and other staples are between 10 and 60 percent higher on the small farms."

The study in Guatemala agrees with the findings of research undertaken in 1974 and 1975 by the Center for the Biology of Natural Systems at Washington University, St. Louis, Missouri. This study, "A Comparison of Organic and Conventional Farms in the Corn Belt," involved 16 pairs of farms located in Illinois, Iowa, Nebraska, Missouri, and Minnesota. Each pair consisted of one organic farm and one conventional farm. The farms comprising a pair were selected to be similar in soil characteristics, size, and livestock inventories.

Observed William Lockeretz, the principal investigator in the CBNS research:

"While it is commonly believed the 'organic farming' is applicable only to small-scale holdings, in fact organic farms in this study are fully commercial operations of a size typical of Corn Belt family farms—

ranging from about 200 to about 800 acres total land per farm. Also, despite another widespread impression to the contrary, these farms all make use of many modern technological advances, except, of course, for fertilizers and pesticides. Modern machinery such as large tractors and combines with corn heads were found on both groups of farms in our study with almost identical frequency. All the organic farms in this study compensated for their non-use of fertilizers by either using livestock manures or by using crop rotations involving soil-improving crops such as clover and other legumes. Most commonly, a combination of these fertilization methods was used. The main crops raised on these farms are field corn, soybeans, wheat, oats, and hay. In this respect, the organic farms are similar to the majority of Corn Belt farms, although their proportion of land in row crops (corn and soybeans) is somewhat lower because of their rotation requirements."

The basic results of the comparative study may be summarized as follows:

Value of crops produced: The value of production per acre of cropland was $165 on the organic farms, compared to $179 on the conventional farms.

Operating costs: The operating costs on the organic farms were an average of $16 per acre lower than on the conventional farms, with most of the differences attributable to the use of fertilizers and pesticides by the latter. (Organic farm costs were $31 per acre, compared to $47 on conventional farms.)

Crop production returns: The organic and conventional farms were essentially identical since the differences between crop value and operating costs offset each other.

Energy intensiveness: The conventional farm sample used an average of almost three times as much energy as the organic farm sample to produce $1 worth of crops. Most of the difference arises from the energy required to manufacture fertilizers, especially nitrogen.

In a concluding observation, Dr. Lockeretz reports: "While our study in no sense provides a basis for a large-scale shift to organic methods, it points out the possibility that a satisfactory alternative may exist that will become increasingly attractive if the problems of energy and fertilizer price increases and shortages become even more severe in the future."

Nations all over the world are in desperate need for food policies that save energy. In this article, we have tried to delineate the diverse aspects of such policies. They include greater use of shorter, direct marketing routes between producer and consumer, all kinds of soil-conserving techniques, legume plantings, crop rotations, more encouragement of home and community gardening, more consumption of

Editor's Note: *The following hypothetical chart, prepared by the Cooperative Extension Service of Pennsylvania State University, shows the amount of money Pennsylvania would have to spend in trucking and fuel if it had to ship in these foodstuffs instead of growing them within the state.*

Estimated Transportation and Energy Costs to Transport Comparable Volume of Food from Primary Point of Alternative Production to Harrisburg, PA.

Food	Volume[1] (in tons)	Point of Production/Processing and Mileage	Trucking Cost[2]	Energy[3] 10⁹ Btu	Energy[3] Diesel Fuel
Milk & Dairy Products	992,247	Madison, WI — 779	$23,168,804	1,855	$6,687,955
Eggs	111,061	Gainesville, GA — 655	2,182,366	174	629,417
Broilers	56,872	Gainesville, GA — 655	1,117,542	89	322,311
Turkeys	9,153	Minneapolis, MN — 1,044	286,692	23	82,679
Apples & Products	187,280	Niagara Falls, NY — 298	1,674,284	134	482,885
Peaches	42,765	Spartansburg, SC — 533	683,817	54	197,220
Potatoes	75,500	Presque Isle, MA — 754	1,707,809	136	492,554
Beef (carcass)	50,699	Iowa City, IO — 840	1,277,623	102	368,480
Pork (carcass)	32,321	Indianapolis, IN — 534	517,786	41	149,335
TOTAL			$32,616,723	2,608	$9,422,836

[1] Volume includes the food production of Adams, Cumberland, Dauphin, Franklin, Lancaster, Lebanon, Perry, and York counties.
[2] Assumes: 40,000 lbs. capacity @ 60¢ per mile one way only.
[3] Assumes Btu per ton mile for diesel tractor is 2,400; a gallon of diesel fuel = 138,690 Btu's; fuel cost = 50¢/gal.

unprocessed foods, and maximum utilization of organic wastes to reduce chemical fertilizer applications.

As is the case with truly effective policies, the individual components are related to others. Thus, besides saving energy directly, using urban wastes in agriculture also encourages the existence of farms in fairly close proximity to metropolitan areas so transportation costs are minimized. The result is a "new food chain" and a more self-sufficient one for states and regions, with a shortened food haul from farm to city and a shortened waste haul from city to farm. As city people begin to garden more, they learn to appreciate the value of fresh, whole foods; they learn about compost and begin to place greater value on so-called wastes. As they value wastes in their own cities and develop uses in gardens, parks, and municipally owned land, then it's not far-fetched to anticipate acceptance of those urban wastes by farmers and rural residents.

Despite what industry public relations spokesmen declare, surveys of Americans show that six out of ten are more concerned with improving the environment than they are with tax reduction or a curb on prices. And obviously, even more Americans would opt for food policies that save energy, and offer more employment opportunities in farming and the food distribution network.

In bits and pieces, we are seeing the gradual adoption of more rational food producing methods and food consumption patterns. But, to achieve real progress toward an energy-saving food system, millions more Americans will have to work toward that goal.

The Greening of the Green Revolution

•

Robert Rodale

Editor's Note: The Green Revolution, which needs chemical fertilizers and pesticides to work, has received a great deal of criticism in recent years. Evidence is growing that such heavy use of chemicals on the soil fails to replenish its nutrients, and also harms inhabitants of the environment which are vital to complete the food cycle. However, scientists who are developing the Green Revolution are becoming more sensitive to these problems and are beginning to respond to them in their research. While serious problems and doubts about the benefits of the Green Revolution remain, the growing awareness of its researchers is encouraging, as the following article reveals.

I first saw the Green Revolution in action when I was in India, about a year ago. Many farmers there were growing the high-yielding types of rice created by the International Rice Research Institute in the Philippines. Those strains of so-called "miracle rice" yield two or three times as much as traditional rice plants. To get that added yield they need more fertilizer, have to be irrigated more efficiently, and farmers also have to spray more with pesticides to protect their investment in fertilizer and water "inputs."

What did the Indian rice farmers think of that situation? Most that I talked to were happy. The energy crisis had already escalated their fertilizer costs, but if they could get irrigation water they were getting good yields and were making more money than they used to.

The Green Revolution has been a very controversial feature of the world food scene, though. The new plants and advanced growing methods have had side effects. Farmers in very poor countries have been converted by the Green Revolution into buyers of chemical fertilizers and pesticides on a rather large scale. They have been made vulnerable to the rapidly rising costs of those products. The varied native strains of wheat, rice, corn, and other plants have been made obsolete—replaced by a handful of super plants. While those new plants produce well for a time, they become vulnerable to epidemics of plant disease. Theoretically, at least, a new strain of a plant virus or other disease could sweep across all of southern Asia, attacking the same plants everywhere. Before the Green Revolution, such plagues were limited because each province or village grew slightly different kinds of plants.

There have been many other criticisms of the Green Revolution. All have stemmed from the fact that the Green Revolution has been an effort to transplant to developing countries an agricultural system that has been developed to suit the needs of America, the richest and most advanced farming country in the world. Our farms are usually large, with few workers. Farms in less affluent countries tend to be very small, and often have a surplus of labor. U.S. farmers have the money or the credit to buy large machines and to finance crop storage. In Green Revolution countries, many farmers are desperately poor. All kinds of other differences in situation stem from those conditions. In the face of those truths, does it make sense to try to superimpose the American system on places like India, Mexico, Thailand, and similar countries?

Those questions were in my mind when I visited the International Rice Research Institute in February 1976. Located at Los Banos, a couple of hours drive south of Manila, IRRI (pronounced like Lake Erie) is the second oldest of the nine international institutes that are centers of Green Revolution technology. And since rice is the basic food of about a third of the world's people, IRRI is probably the most important. Founded in 1962 with Rockefeller Foundation money, IRRI now gets much help from other foundations and especially from the U.S. government and the governments of Asian countries. It's current budget is about $10 million a year.

From the beginning, IRRI has tried to do things in a first-class way. Robert F. Chandler, Jr., the original director, was a stickler for perfection. One staffer told me how he fussed until even the memo forms were printed just right. Chandler is now semi-retired and lives in Massachusetts. The current director, Nyle C. Brady, is one of the great names of U.S. agricultural science. His book *The Nature and Properties of Soils* is a classic text, known and used by hundreds of thousands of agricultural students. Before coming to IRRI, he was Director of Science and Education for the U.S. Department of Agriculture.

Many of the people I met spoke of contentment with their work. What pleased them most was the spirit of international cooperation and the lack of commercial secrecy and competition. Dr. W. Ronnie Cauffman, a senior plant breeder, told me that at IRRI he could get his hands on any strains of rice he wanted. "Back in the United States," he commented, "much of the crop germ plasm is in the hands of companies working on their own special strains. A breeder can work with only part of what's available." Several scientists told me that they preferred to spend their whole careers at the international institutes, where they could get the cooperation needed to do good work.

When IRRI was formed, farmers throughout Asia were growing types of rice that had long, weak stems and wide, droopy leaves. Fertilizing produced heavy heads, which made the plants lodge, or fall over. The grain would then get wet, or be eaten by rats. Yield was low.

IRRI scientists saw the opportunity to produce semidwarf strains of rice, which would have strong, stiff stems. Within a few years of the founding of the institute, they did just that. Their new rice strains not only could stand up well when fertilized, but also had narrow, erect leaves which allowed more sunlight to penetrate the leaf canopy. That increased photosynthesis, in effect making the rice plant into a more efficient collector of solar energy.

The most famous of these early improved strains was IR8, the first "miracle rice." When handled according to IRRI directions, IR8 and some of the other early strains yielded three to five times as much grain as traditional varieties. The Green Revolution in rice farming was given its first push forward by that plant.

Then the bad reviews started coming in. IR8 didn't taste as good as the old rice. Its grains tended to fall apart and get mushy during cooking. Insects and disease attacked the plants. Many farmers weren't able to supply the fertilizer, pesticides, or extra irrigation water needed to get large yields. Big farmers had more success with the new rice strains than the poorer farmers, creating social problems. The "miracle" got a black eye.

The troubles with IR8 were not entirely unexpected. "They knew IR8 had weaknesses when it was introduced," Dr. Brady told me. "But those early rice strains gave us a breathing spell. They were trying to keep food production ahead of population growth. IR8 showed what could be done."

IRRI has learned from its mistakes and is aiming its efforts in very practical directions. "Today we work more in the real world—not in the world of experiment stations and rice farmers," Dr. Brady continued. Helping the small farmer is the watchword. Small farmers, after all, is what Asia has by the millions. If they are displaced from the land, there will be no room for them in the already jam-packed cities.

Dr. Brady told me that the current IRRI thrust is to create rice plants that need fewer inputs. Built-in tolerance to pest attack is being given very high priority in the development of new rice strains. Ability to withstand floods and droughts is also an important goal of the plant breeders. The need for fertilizers is being reduced by the design of new methods for placing nutrients as close as possible to the root zone, where they won't wash away and where plants can reach them easily. Much research is going into the development of total cropping systems, blending the new plants, improved small machines and all other ideas for improving production into a total package. Almost all the research on these cropping systems is done on farmer's fields, instead of on experiment station plots. Scientists find quickly what problems the small farmers have, and can try to solve them. The farmers themselves are deeply involved in the process of creating new cropping systems.

IRRI's biggest recent breakthrough was revealed to me by Dr. Cauffman, a quiet-talking Kentuckian. "We now have a rice that is so resistant to insects that it can be grown without the use of any pesticides," he said. "It's IR36, and will be released this fall. We've tested it at seven different places in the Philippines."

I stopped taking notes and looked him in the eye. "Tell me that again," I asked.

Ronnie Cauffman leaned back in his chair and gave me a short lecture on rice insects and the problems of pesticide use. "The majority of Asian farmers can't afford pesticides," he began. "Not only that, but some of the worst insect pests can't be controlled by poisons. The brown plant hopper is one. The brown plant hopper not only causes hopper burn, but it transmits grassy stunt virus. Tungro, the worst virus disease of rice, is transmitted by the green leaf hopper."

"If you spray to try to control these insects, you're talking about five different pesticide applications," he said. "That's quite expensive. The Japanese and Taiwanese have emphasized the chemical approach to rice insects, because they can afford it. But they're finding that chemicals really don't do the job. Actually, the poverty of the farmers we are working with may be a stroke of good fortune. It's forced us to develop resistant plants, which really are a solution."

"What do the chemical companies think about your work?" I asked.

"We have those fellows worried," Dr. Cauffman said. "They turn white when we bring them in here and show them rice plants resistant to darn near anything. You name the rice problem today, and we have a variety that's resistant to it."

My next question was whether insect and disease resistance like that could be bred into crop plants grown in the United States.

"Theoretically it's possible," Dr. Cauffman explained to me. "But from a practical point of view it would be very difficult. There's so much secrecy in plant breeding in the United States. You don't have access to all the germ plasm the way we do here. There'd be too much frustration."

Had we talked longer, we might have mused about how the pesticide approach is ingrained into the thinking of most American farmers. They are comfortable with the idea of trying to kill bugs. Most farmers in the United States just don't realize that it's possible to create plants that can stand up to almost any kind of insect attack. A massive research effort would be needed to do that, but it is possible.

On my last day at IRRI, I spent a final half hour with Director Nyle Brady. We talked mostly about the history of IRRI, and its changing mission. Finally, I asked him what he thought of organic farming.

"I went to China last year and saw how they use every bit of organic matter," he started. "The Chinese system of recycling is fantastic. I wish the same principles could be used in the Philippines, and elsewhere. It's a sound practice—no question of it."

"Back in the days when nitrogen fertilizer was cheap," Dr. Brady continued, we used to use 150 kilograms per hectare (133 pounds per acre) without worrying. But the price of fertilizer now is forcing us to use techniques that will let us grow more food with fewer inputs. It makes me sick when I see rice straw being burned—the common practice throughout Asia. They haven't yet learned how to make compost."

After returning home, I mailed to Dr. Brady a copy of Dr. F. H. King's book *Farmers of 40 Centuries*, which describes in great detail the Chinese farmer's passion for saving and using organic matter. Dr. King was a high research official in the U.S. Department of Agriculture in the late 1800s, who spent several years in the Orient observing farming methods. I can think of no book that tells more convincingly the need for organic practices in Asia than does *Farmers of 40 Centuries*. In a letter of thanks for sending the book, Dr. Brady commented that "The messages contained therein are not only interesting, but of vital importance."

Clearly, the Green Revolution I saw being created now by the International Rice Research Institute is quite a different thing than the Green Revolution I had read about. Many of the mistakes of the past have been corrected. New work is definitely headed toward making the life of small farmers easier. There are important organic directions to the work being done, though it is not a plan for organic farming the way we think of it here. But in the Philippines and much of Asia, the Green Revolution is getting greener.

The Return of the Draft Horse

•

Maurice Telleen

We are now entering the fourth decade since the power structure of American Agriculture (since renamed Agri-business) decreed that the draft horse and mule had become useless artifacts of history. In that respect the horse has been in some excellent company, such as crop rotation, small owner-operated farms, diversified livestock and crop programs, home gardening, the use of natural fertilizers (manure), and thrift. In brief, the decline hit most of the things that had lent stability to agriculture in good times and endurability in bad times ever since the days when Tom Jefferson had envisioned his race of farmer-philosophers, which would serve as the balance wheel and strength of the republic. All of these things were discarded in haste following World War II. The work horse and mule were but two of many casualties.

The fact that today, some 40 years later, the demand for draft horses and mules far exceeds the available supply suggests that those judgements handed out so freely in the late 40s were both wrong-headed and short-sighted.

Our country was full of farmers who regarded them in just that light, but the overwhelming weight of official, social, and economic opinion took its toll and eventually most of them discarded their last teams, quit farming, or died off. Thousands of farmers who abandoned the horse at that time did so with nothing more articulate than "a feeling in their bones" that this was wrong, just as they adopted the blanket use of

Besides farming with horses and raising Oxford sheep, Maurice Telleen is Editor of The Draft Horse Journal *(Route 3, Waverly, Iowa 50677), a quarterly publication devoted to raising and using heavy horses.*

chemicals for weed and pest control. They had few champions of their view to confirm their beliefs in public or in print, so ultimately they were intimidated into accepting the unbelievably vain concept that the "conquest" of physical nature was an accomplished fact and that we had entered a new era in terms of our relationship with our earth.

In this sort of social and economic climate the draft horse and mule stocks of this country dwindled to the point of invisibility to most people by 1960. But three basic groups were responsible for keeping a nucleus of draft breeding stock alive through those times. Most obvious were the purebred breeders who kept and propagated heavy horses for show purposes and the big hitches. For the most part, this group was, and is, made up of prosperous "agri-businessmen" who kept the horse, not because they believed in him as a source of power, but simply because they liked him, enjoyed showing him, and in many cases were carrying on a family tradition of horse breeding.

The second group, considerably larger numerically, were the Amish who retained the horse as their only source of motive power and as an integral part of their way of life, both economic and cultural.

The third group, and the one we are dangerously short of (40 years does take its toll) are non-Amish farmers who keep horses, frequently in combination with tractor power, out of a reasoned belief that they are the most economical source of power, a keystone in maintaining soil fertility, and fundamental to the economic independence and continuity of the small family farm.

Many of the farmers in this latter category are either wholly or predominantly organic. This is probably best explained by the fact that horse farming *does not* lend itself to: 1. one crop agriculture (You must have pasture and hayland for the work stock, so you almost always find other types of livestock and crop rotation being practiced on such farms.); 2. huge acreage (You simply haven't the time to work it.); and 3. farming distant patches of ground. You don't take four horses and a disk 10 miles down the road. You farm, in effect, within sight and easy walking distance of your home. Whatever the reasons, there does seem to be a natural marriage of thought between organic agriculture and a source of power that reproduces itself, with good care is self-repairing, consumes homegrown fuel, and contributes to the fertility of the soil he tills with his own manure. Horse farming and organic farming are very comfortable with one another.

Without belaboring the economic and ecological ramifications of horsepower, I would like to quote one of those "reluctant converts" to mechanization. The following statement appeared in the *Suffolk Bulletin* of 1946. The writer was Lloyd Wescott, owner of Mulhocaway Farms near Clinton, New Jersey, a man who was in the forefront of Suffolk horse breeders at that time. It was written in response to an article in *Farm Journal*, wherein True Morse of Doane's Agricultural Services had advised farmer-readers to "Sell that last team." Mr. Wescott's response, in part, follows:

"As I gave up the use of horses with greatest reluctance, I have examined my reasons for so doing very carefully and feel I can give all the answers as to why they cannot be used with some authority. But there are still serious questions in my mind.

"Consider, first, the investment in power machinery as compared to horse-drawn machinery. Last winter, when offering to sell some of my useless equipment, I went back through the records to determine what I had paid for some of it. It now costs about as much for a single tire for a tractor as it cost for a sulky corn cultivator about eight years ago. One could then equip an entire farm for what a tractor and cultivator now cost. Can farming support such an investment?

"Consider the size of farm needed to utilize the larger power units such as the field harvester, the pickup baler, and the combine. I find one of each adequate for the operation of my large unit. Custom work has proven unsatisfactory in most cases. Can the American farmer work out a plan for joint neighbor ownership of this equipment or does this spell the end of the family-sized farm?

"Consider the cost of upkeep. Mr. Morse gives figures for the cost of keeping a team, the income from the extra cows, and cost of operating a tractor. Will the income from the cow stay where it is now? Can we overlook the fact that the cost of keeping a team is largely money paid back to oneself for hay, oats, and labor, while the cost of operating a tractor is cash out of pocket for gas, oil, repairs, and replacement?

"Consider also the broad economic aspects of the problem. Should the day come when the farmer is again faced with ruinous surpluses, will these not be much greater than they were when the acres that went to feeding horses will grow crops to sell? And what of our dwindling natural resources of petroleum? Will we eventually raise crops that are sold to be processed into fuel, to be repurchased by us to burn in our tractors, where we now have available a hay-burner of our own?

"Under present conditions, I cannot afford to work horses. But the change from horse to tractor farming is a profound change. As a result the farmer will lose a measure of his independence; his fate will be more closely linked to the strength and effectiveness of organized labor; the family-sized farm may be the next aspect of rural life to be found obsolete and uneconomical."

Most of the concerns voiced by Lloyd Wescott in 1946 have become the realities of the 70s. Agriculture was full of such reluctant converts, few as articulate as he, but all sharing the common fate of having their doubts and their thoughtfulness being dismissed as useless nostalgia by the movers and shakers of the brave new world after World War II.

So much for background on the decline of the work horse. Now for a look at his recent resurgence.

About 1960 the draft horse market began to strengthen, ever so slightly at first, and with great acceleration in recent years. The reasons are manifold. Firstly, our work stock inventory had gotten so low that those wishing to use them simply had to ante up a little more, a case of supply and demand. The wretched prices commanded by good young work stock in the 50s had brought breeding to an almost complete standstill. Second, enough time had elapsed from the immediate post-war period when farm youngsters were ashamed of fathers who hung onto a horse so that the "stigma" of using horses was past, at least in part. Third, it became sort of fashionable in some quarters to have a team around. Fourth, the show and parade fad took hold a bit more. Many fairs which had dropped draft horse classifications in their rush to appear "modern" in the post-war period discovered that things had become so modern that they were almost useless in their traditional roles as agricultural institutions. Consequently, they turned more and more to entertainment and to attractions such as the big horse hitches. At his best, the draft horse is a spectacular beast, and much more exciting to the urban crowds our fairs found themselves playing to, than a calf or lamb—however good and useful they might be. The same can be said of parades. A team of horses was no big deal to the youngsters of the country when True Morse was urging everyone to send them to the kill, but they are to the sons and daughters of that generation. So, rather suddenly, there was a superabundance of places to "show off," and it stimulated the market. Fifth, any rising market attracts buyers, just by being a rising market. Farmers who hadn't owned a horse for a decade suddenly found themselves back in business. Lastly, and most importantly in the long pull, people began to raise serious questions about the dogmas of agri-business. The result has been a rediscovery of the horse in his traditional role as servant, partner, and preserver on the land.

I don't pretend to know what the future holds for the heavy horse. What I am certain of is that for those who wish to farm a modest amount of ground with a minimum of cash outlay and do it with a source of power that is self-reproducing (colts to sell instead of depreciation), consumes homegrown fuel, and contributes to the fertility of the soil, the draft horse can and should have a place in their future.

There are some obstacles, of course. Horse machinery, a few years ago both cheap and plentiful, is now neither. Rust and the cutting torch of the junk men have taken a heavy toll. Good pieces now command respectable prices when you find them, but still far under the investment required for tractor farming. Horse farmers lost their support industries. Hopefully, this trend is being reversed. An example is the firm of F. A. Hochstetler and Sons (Rt. 2, Box 162, Topeka, Indiana 46571), where they make new horse-drawn farm wagons and sulky plows. Last year they manufactured and sold 174 new horse-drawn plows—not a great number in the land of mass production, but a possible indication of things to come. We hope so.

Harness making lends itself to the cottage industry approach even more readily than the manufacture of plows, mowers, and other machinery. New harness is now available at many shops throughout the Cornbelt states.

Along with good machinery, there's a dearth of experienced teamsters able to teach draft horse skills. Obviously a lot of the best practitioners of the art of horse farming are now in the cemeteries and quite out of reach. Treasure the ones that are left and willing to help you! If you will but look around, most rural communities still have some good old horsemen who will be happy to "come out of retirement"—not to do it for you, but to show you how.

If you decide to go out and get yourself a good broke team, try to take an experienced horseman with you, pick up the pieces of usable horse machinery you can find at local farm sales, prepare to work a little harder physically than your mechanized neighbor, and go at it kind of easy, not all in one jump. Work a pair for a while on the spreader, corn planter, and seeder before you put four on the disk, and you may just discover that horse farming is what fits you, your concept of what farming should be, and your place on this earth.

Farming With Horses

•

Wendell Berry

In the work itself, horses have certain advantages over tractors. They can be used safely on steep ground where a tractor would be either dangerous or useless. A horse farmer can get into his fields more quickly after rain than can a tractor farmer. And horses do not pack the ground as much as tractors. It is generally acknowledged among the tobacco growers of Kentucky that the work of horse-drawn cultivating plows has never been equalled by any tractor.

Beyond these practicalities, there is the satisfaction that one gets from working a good team. A tractor may be handy, always ready to use, untiring, enormously powerful; but it is not alive, and that is a great difference. To work all day with a well-broke and willing team is a pleasure as well as a job; it is a cooperative venture, a sort of social event. And when the day's work is finished, to stable the team and water and feed them well, or to turn them out onto good pasture, is a comfort and a fulfillment. Between a farmer and a team there exists a sort of fellow feeling that is impossible between a farmer and a tractor, and for me that rates as a considerable advantage.

Another thing I like about working with horses is their quietness. When you work with a tractor you hear nothing but the tractor; it is a kind of isolation. The engine roaring in your ears all day, you lose awareness of the other life that is going on around you. With horses, unless you are using some noisy implement like a mowing machine,

---

Wendell Berry is a poet, essayist, and Distinguished Professor of English Literature at the University of Kentucky. He also homesteads, and uses horses on his farm.

---

you hear the wind blowing and the birds singing and all the rest of the stirrings and the goings on of the countryside.

And so there is a good deal that can be said in favor of farming with horses. There are, in fact, a few widely scattered farmers who still do farm exclusively with horses. There are many more who keep a team or two for part-time use. And of course there are the Amish, for whom farming with horses is a community ideal and a way of life. Nevertheless it would be irresponsible simply to recommend to anybody who owns a farm that he should sell his tractor and buy a team.

Aside from the difficulty of locating a team and equipment, it is important to realize two things before taking on horse farming.

1. Horses are not standard. No two are exactly alike in looks, size, conformation, or disposition. It is therefore extremely unwise for an inexperienced person to attempt to buy a team on his own. I have said that a good team is a source of pleasure. It is now time to say that a bad one is a curse—a nuisance, a liability—sometimes, a danger. Some horsemen will take pride in selling good stock and in satisfying their customers. Some will sell poor stock with extravagant praise.

2. Most important: *It is not easy* to work and care for a team. It cannot be learned easily or quickly. A person who can drive a car can probably teach himself to drive a tractor in a short while. To learn to drive a team well, you need a teacher, and you need experience. An inexperienced teamster can easily injure or kill his team. And can easily get injured or killed himself.

Obviously, then, a person inexperienced with horses who wants to farm with them is much in need of the advice and instruction of an experienced teamster whose intelligence and judgment can be trusted. And this brings us to the final difficulty: such people are getting scarce. The last generations that grew up working horses are dying out, and their knowledge is dying with them.

Fortunately, there are some experienced horsemen scattered over the country who have had these problems on their minds. And because of their efforts there begins to be some promise of help for the would-be or the novice teamster.

First, I want to mention, and recommend, *The Draft Horse Journal* (Rt. 3, Waverly, Iowa 50677), edited by Maurice and Jeannine Telleen. This magazine is an assembly point for all sorts of information about sales, breeders and dealers, suppliers of equipment, etc. There are articles on the history, breeding, and care of draft horses. And in the short time that I have been a subscriber, there has been an increasing number of articles on the use of horses for farm work. Anyone interested in farming with horses would find this magazine both a pleasure and a valuable investment.

In the Summer 1973 issue of the *Journal,* I was much interested to see an article which announced the first session of a Teamster's School to be held at Indian Summer Farm, Cabot, Vermont, September 23–27. Ted Bermingham, according to the article, had undertaken to set up the school on his farm with the hope of preserving and passing on the knowledge and skill of "the last full generation of working teamsters." For this purpose Mr. Bermingham had rounded up a faculty of experienced horsemen to teach the disciplines of farming and logging with horses.

Mr. Bermingham was kind enough to permit me to visit and observe the school. I arrived at Indian Summer Farm in time for the night session on Wednesday, the 26th of September. I stayed on through the school's concluding sessions on the next day, getting informed about the workings of the school, talking with students and teachers, riding along on various work and pleasure jaunts, and in general having a lot of fun.

Indian Summer is a most hospitable place high on a mountainside above the village of Cabot, looking out across some of the loveliest country I have ever seen. The farm has a large indoor arena, plenty of open land and woodland, and a good many generous neighbors interested in farming and logging with horses—all of which fits it admirably to the needs of a teamster's school. Ted and Norma Bermingham have the interest and energy necessary to their enterprise, and they are excellent welcomers.

The Economics of Horses vs. Tractors

"An Illinois farmer who owns a $9,00 diesel tractor recently compared his costs of operation with that for his Belgian horses.

"His tractor burns 40 gallons of diesel fuel a day—$6.40 plus the cost of maintenance and the cost of depreciation which will reduce the value of the tractor to very little within 10 to 15 years.

"But a working horse does fine on 17 cents' worth of oats and 18 cents' worth of hay a day, plus night foraging in the pasture. He supplies 10 tons of fertilizer a year for the land. Before he grows too old to work, he'll produce his own replacement, plus others for resale. He can do a full day's work for up to 16 years."

—*THE DRAFT HORSE JOURNAL*
Autumn 1973 issue

The school was staffed by three teamsters: Harrison Miles, Sidney Lowrey, and Archie McCarty. There were also a number of "visiting faculty": other teamsters, loggers, a horse dealer, a geneticist, a veterinarian, etc. And there were 20 students who came from 11 states:

Colorado, New York, Connecticut, Massachusetts, Michigan, Maryland, Maine, Indiana, Vermont, New Hampshire, and New Jersey.

Tuition was $150. This paid for all instruction, stall and board for a horse or team (if any), supper on Sunday night, and three meals a day, Monday through Thursday. Daytime instruction involved working at the chores of farming and logging. Study of the various tools and techniques always took place amid the practicalities of actual use. Students drove their own horses, or horses furnished by the school, under close supervision of the teachers. They also took part in the necessary stable work. The night sessions consisted of talks on such topics as horse husbandry, logging techniques, breeding, equine genetics, veterinary medicine, horse trading, and shoeing.

Mr. Bermingham and his staff are well aware, of course, that the teamster's art is complex and dependent upon experience, and that no novice could hope to become proficient at it in a few days. The school is meant to offer as full an introduction as possible to skills that the beginner can then practice on his own. There is no doubt that a few hours of work with an experienced teacher can save many hours of expensive and dangerous self-instruction by the method of trial and error. The students I talked to all believed that the school had been thoroughly worthwhile. So did their teachers. And so did I.

I hope that such schools will soon be undertaken in other parts of the country. They would be good for the draft horse business, and they would be good for farming.

The school was sponsored by the Draft Horse Institute, established in 1972 by Ted Bermingham, to collect and maintain the working knowledge of farming and using draft horses before it becomes lost. For information on the future school planned for this coming fall, you can write to Mr. Bermingham at Indian Summer Farm, Cabot, Vermont.

Part III

•

Power For and By the People

Generally speaking, the closer one is to an energy source, the more efficient is the use of that energy. And in some instances, the most efficient means of energy production and conservation (which makes the energy we *do* produce go that much further) are literally in our own backyards—or kitchens or living rooms.

Efficient and widespread use of solar energy for heating and cooling buildings and supplying electricity is not here yet, but the sun can be used for simpler purposes—like providing heat and light for a home greenhouse to help grow plants and also to warm up adjacent living areas. Buying cords of wood to burn in a centralized furnace is impractical, but wood from one's own woodlot, burned in efficient fireplaces and woodburning stoves for heating and cooking, can make good sense.

Recycling as a means of conserving energy resources is a fine idea, but on a large scale it hasn't been very successful. There's just not enough money in it to interest investors and government officials. Turning kitchen wastes into usable compost, however, is a way every homeowner can get something from "nothing." The larger the scale of our food production, the more problems we encounter; soil management, pest control, harvesting methods, and transportation cross-country can be headaches for agri-business. On the other hand, when the food you eat comes out of your own backyard *organic* garden, you are involved in the most efficient means of food production possible.

A Solar Greenhouse for All Seasons

•

Steve Smyser

While some gardeners are spending $100 a month to heat their greenhouses this winter, Christopher and Melissa Fried are using their greenhouse to help them hold their total winter heating bill to about $30.

Besides producing a quarter of the heat required to warm their nine-room farm home near Elysburg, Pennsylvania, the Frieds' homemade "sun structure" keeps them in fresh vegetables year-round and serves as a pleasant place to relax in a light, summery atmosphere in the dead of winter.

When Fried packed in his aerospace engineering job in Long Island for an uncertain future on a nine-acre farmstead two years ago, one of his primary objectives was to liberate his family from dependence on the oil companies and utility giants, along with the air and water pollution they help create. His other goal was to acquire the means of producing a steady, year-round supply of inexpensive, high-quality food for the family.

The attractive 24-foot by 16-foot by 14-foot lean-to greenhouse that now graces the south side of his two-story house answers both requirements admirably.

At the going prices of oil, gas, and electricity, Fried estimates it would take at least $500 a year to heat his home. By plugging into free energy from the sun—primarily in the form of solar radiation captured and put to work in the greenhouse and secondarily in the form of stored solar energy (wood) burned in a high-efficiency wood heater that serves

Steve Smyser is an Associate Editor of Organic Gardening and Farming and a Contributing Editor to Environment Action Bulletin.

as backup to the greenhouse—Fried's net outlay is for the electricity to run the ½-h.p. blower and the gasoline for his chain saw.

The operation of the system is simplicity itself. As solar radiation enters the 350-square-foot fiberglass sun wall, it strikes the ground and walls inside and quickly warms the sun room. When the temperature reaches 95 degrees F. (around 11 A.M. on a clear day in winter, when the sun's rays strike the earth at their lowest angle), Fried opens the kitchen and living room windows (both contained inside the solar system) and a thermostatically controlled fan blows the hot, moist air inside. Late in the afternoon, or when cloud cover develops and the inside temperature drops below 75 degrees F., the blower goes off and the windows are closed.

The wall of the lean-to, extending 12 feet out from the house at its base and slanted at an angle of 56 degrees (optimal for the collection of solar radiation at Fried's latitude), consists of two layers of Kalwall Sunlite fiberglass (.025 inch thick) fastened to 1 x 6 rafters on 2-foot centers. The upper two-thirds of the wall (250 square feet) covers aluminum, air-type absorbers, while the lower third remains unblocked so that light will strike the plants on the ground. The house cellar door and two windows are included inside the structure for an entrance and air circulation. Double-glazed windows on each end of the structure allow for the entrance of compost, rotary tiller, summer ventilation, and early and late sunlight.

The absorbers Chris made from recycled printing-press plates are inexpensive ($20 total) yet remarkably efficient. Fabrication involved bending them so that fins to enhance hot-air movement protrude on the back side, then painting them flat black. Air is circulated throughout the system with a ½-h.p. fan that blows it through 1-by-2-foot ducts.

Heat storage, when completed, will be by rocks in the cellar. In the area previously occupied by the coal bin, Fried is building an insulated bin that will hold about 25 tons of rocks.

By using his cabinetmaking skills to keep waste to a minimum, and scavenging for used materials where possible, Fried has kept the cost of the entire system to about $650, half of which went for the Kalwall. Various solar devices that could have improved the performance of the unit were considered but ultimately rejected as adding too much to the cost and complexity of the intentionally simple design.

With subzero temperatures fairly common and at least partial cloud cover four days out of five in his part of central Pennsylvania, Fried knew he'd need a heavy-duty backup system if he wanted to completely avoid oil or coal. On a clear day, the sun room can generate 50,000 Btu's an hour (the capacity of a small conventional furnace); Chris wanted his backup heater to have a capacity at least equal to this.

The solution was a homemade woodburning heater of the downdraft, forced-air type. The design is basically a marriage of what Fried considers the best features of the Ken Kern drum stove (September 1975 Organic Gardening and Farming magazine) with the best of the famous Riteway "airtight" wood furnace. The whole system is thermostatically synchronized so that when temperatures in the greenhouse drop, the windows are closed and the stove takes over.

For air circulation there's a metal cabinet surrounding the drum stove. The old furnace filter/blower unit is suspended from the beams beneath the kitchen floor. Air from the basement flows up through a hole in the floor, into the cabinet, back down through the floor through ducts to the living room and/or bedrooms, and finally back to the blower. One loading of wood lasts about 10 hours.

Growing plants in a variable temperature greenhouse requires an altogether different approach than the traditional (and expensive) practice of simply turning up the heat to meet the requirements of warmth-loving plants. It means adapting your plants to the seasons, not vice-versa.

This past winter, while Fried's rock bin heat storage system was still in the planning stage, he stuck with crops that could take the cold nights—lettuce, spinach, endive, carrots, parsley, etc. He is confident that by making full use of his unit's vertical space and planting on a staggered schedule, the room will supply all the vegetables his family can use.

While solidly optimistic about the economic competitiveness of new homes specifically designed for solar heating (he's planning to build one of his own in two or three years), Fried has reservations about the cost effectiveness of retrofitting existing homes to active solar systems.

"Nature," he feels, "provides us with plenty of clean, free energy, if we would but appreciate and fully utilize what we have." Chris has found that living so intimately with the subtleties of the sun produces a heightened sensitivity to the energy "personality" of one's home. Besides the obvious conservation measures like thorough insulation, weatherstripping and sealing off rooms that get only seasonal use, there are dozens of small techniques and habits of energy consumption that added together can improve energy efficiency by a quarter or a third in even the most out-of-square old farmhouse or shoddily constructed tract home.

While most of the energy planners and government agencies continue to be obsessed with ways of producing even more energy, Fried is concentrating on helping others in his community find greater energy independence by reducing consumption and doing more with less. Besides serving as a living example of how a low-energy lifestyle can improve the quality of life, he teaches a continuing education course at nearby Bloomsburg State College on "Energy Conservation in the

Home." Participants are characteristically amazed that such substantial energy savings can be realized by relatively minor changes in daily habits.

For the time being at least, the Frieds are living the healthy, self-reliant life close to nature that they had envisioned when they first moved to their farm. Even in their remote valley, however, far from urban pollution and population pressures, the spectre of Big Energy is rearing its menacing head in the form of a giant Energy Park (euphemism for nightmarish concentrations of nuclear plants in rural areas where utility companies anticipate the least organized resistance) planned for a 60-square-mile site about 25 miles distant. Knowing firsthand the needlessness of such destructive solutions to energy gluttony, Fried is deeply involved in the effort to stop the project, speaking to interested groups, contacting public officials, and helping to coordinate an organized opposition at the hearings being conducted on the as-yet-unresolved plan.

By symbolizing their low-impact lifestyle and embodying their convictions about how the earth's resources can best be used, the Frieds' solar greenhouse is convincing evidence that there is a better answer.

Methane From Manure at the Organic Demonstration Farm

•

Steve Smyser

Manure power is alive and well at the Rodale Organic Demonstration Farm near Maxatawny, Pa. After a month-and-a-half in the works and five anxious weeks of head scratching and trying to project total confidence to all doubters, we're making breakfast over a hot blue jet of methane.

We're not the first to do it, and as high technology goes it may not quite rank with the invention of the internal combustion engine or the nuclear breeder reactor, but then again it might be around long after these more glamorous sources of energy have passed from the scene.

The methane digester is one of the first fruits of a work-study program recently initiated by Rodale Press. This particular project was developed in conjunction with the Environmental Studies Program of the State University of New York at New Paltz.

Essentially nothing more than a closed tank in which organic waste (chicken manure, in our case; vegetable waste works, too) is allowed to decompose in the absence of air, the digester is the perfect "closed system"—from plant to animal to digester to plant. Like the compost pile, it utilizes the decomposition of old life to create and enhance new life. And the pollution-free fuel that is generated as the chief by-product of this process can be used to perform dozens of heating and cooking jobs around the homestead—household heating and cooking, crop drying, heating incubators, farrowing pens, greenhouses and other farm structures, as well as powering an engine or generator to produce electricity.

Every bit as valuable as the gas—at least for the gardener or farmer— is the slurry that is removed from the tank after the "soup" has produced its methane. With a high level of available nitrogen, as well as significant quantities of phosphorus, potassium, and metallic salts, the

97

digested slurry makes an excellent fertilizer and soil-conditioner. We'll learn just how excellent as we begin to study its effect on crops and soils at the farm over the coming months.

John Riewick and Eric Friedman, the SUNY New Paltz students who worked with Farm Manager Richard Weinsteiger on designing and building our pride and joy, kept a logbook of the experience. Here's an excerpted account:

"Our first task was to convert a used 275-gallon oil tank to suit our needs. The tank had to have several hours of welding done on it in order to make the necessary fittings. In addition, the interior of the tank, lined with oil residues, had to be thoroughly cleaned out, and the exterior scraped and painted. To keep the slurry well mixed, an agitator constructed of four paddles screwed into a steel shaft was installed at the center of the digester.

"When using solar energy as a heat source, good insulation is essential. You have to store heat to compensate for the time when the sun is not shining. To do this we constructed a special container for the digester tank. The southern side of this container consists of windows angled at 60 degrees to maximize solar radiation. The rest of the container, including the floor, is insulated with 3.2 inches of polyurethane foam. Polyurethane is the best commercially available insulation; it is twice as good as the same thickness of fiberglass wool. To keep heat from being lost through the glass windows, we used the same amount of insulation to make an adjustable panel which can be easily closed at night or other times when heat losses through the glass exceed the solar heat input. Inside and outside temperatures are monitored by means of an 'indoor-outdoor' thermometer attached to the outside wall of the container.

"The digester tank fits snugly into the container. Filled up with slurry (an equal mixture of wastes and water), the weight of the tank is about one ton. This entire mass, the metal of the tank included, has a fairly high heat retention capacity. Thus, the solar heat is for the most part maintained by the mass of slurry itself. The acceptable temperature variance of the slurry is from 110 to about 65 degrees F. Below 65 degrees F., heat tape around the digester tank can be turned on to supply needed heat, but we don't anticipate needing this external heat source very often.

"The methane rises out of a hose on top of the digester to two holding tanks—50-gallon oil drums with their open ends in water. As the drums fill with gas, they rise in the water.

"On the way to storage, the gas passes through a water filter to remove any condensed water, and through an ash filter to remove carbon dioxide.

"The amount of methane a digester can generate is determined by the

Framed in 2-by-4's and protected from the north by concrete blocks and banked earth, the digester housing was intended to be as attractive as it is functional. Good insulation is critical—we used 3.2 inches of polyurethane foam throughout.

Fifty-gallon drums rise in water as they fill with methane coming in by way of the CO_2 filter. The system is producing gas enough for about three hours of burning time a day.

The digester tank is mounted on caster wheels for easy removal from the housing. Outlet hose carries methane from the digester to holding drums. Tank temperature is monitored by indoor-outdoor thermometer. Hole in door of housing (sealed with inner-tube rubber to prevent heat loss) is for agitator handle.

type and quality of waste being used, the temperature of the tank, and the length of retention time for the slurry. Our gas output is presently averaging around 12 cubic feet (three hours burning time) a day. That works out to around 7,500 Btu's.

"Because the effluent is not exposed to weathering during its digestion time, there is no nutrient loss due to leaching or ammonia evaporation, as there is in aerobic composting. One of the ongoing aspects of this project in the months ahead is to analyze the nutrient content of the slurry and evaluate its effectiveness as a fertilizer on various crops and soils at the farm.

"In all, the digester cost $700 to construct. Now, with what we've learned from building the prototype, we could do it for less," Riewick and Friedman report.

Wind Power for the People

•

Joe Carter

Wind. It comes; it goes. For most of us wind is a commonplace, a passing turbulence, not much to notice save when it grabs a hat or otherwise swoops our way. There is a small and growing group of people, however, for whom wind is an important event: when the wind blows for them it's also making electricity. For a variety of reasons these people have sought the energy of the wind to supply their own energy needs through the use of wind powered generators.

Paul and Sally Taylor's Australian-made "Dunlite" wind generator has been operating for almost a year. The Taylors, along with Baggins the hound, horses, ducks, and cats, live on six acres of gently rising pastureland in an area known as Hessel, a bit east of Sebastopol, California. Two years ago they came to Sonoma County from British Columbia, where Paul was a dentist; now he practices in Cotati.

Coming into Hessel, it is evident that the wind has been working for many years. Old water-pumping windmills like the "Aermotor" can be seen here and there, some near familiar water towers. Partway up the Taylor's drive stands one of these water-pumpers. It's another way they are using the wind. Paul had it installed over a shallow well, an old mill put to use again.

The house is at the top of the rise, and just away from it sits the Dunlite atop a metal tripod-tower some 25 feet high. When a breeze picks up, the three-blade propeller silently starts with it, turning gears that step up the rpm's to the generator behind. At around seven mph wind speed, the generator is turning fast enough to deliver useful d.c.

Joe Carter is a writer on alternative energy in San Francisco, specializing in wind power.

power to the house or to the storage batteries for times of no wind. With wind power, power storage is critical since the raw material is so highly inconsistent. So far, batteries, though not highly efficient, are the best solution for small-scale operations like power for homes. Awaiting further development is a way of using a spinning flywheel to store the wind's energy as kinetic energy and then to be drawn out again for use as electricity.

In a shed beneath the Taylor's wind generator are ten 12-volt truck batteries and the voltage regulator control panel. This setup gives them a normal 120-volt current with considerable storage capacity. Paul figures that, fully charged, the batteries could deliver full power for five to seven days of no wind, depending on the load demanded.

Paul is fairly satisfied with the performance of the Dunlite. "We're doing all our lights and some small appliances on wind power. When it's really blowing outside we can hook in the washer. We're very careful with our electricity." Since the generator doesn't supply enough electricity for all their needs, the Taylor's home is still tied in with PG&E power. But it is the feeling of being producer, manager, and consumer of their own energy, relatively independent of the power giants, that has drawn Paul and Sally to wind power.

The Dunlite is rated to produce 2,000–2,750 watts of power at its peak, which means a wind of 25 mph. With adequate wind and storage capacity, this is enough for an average household's consumption, save for heavy loads like cooking and heating. Proper site selection is vital. Wind generators are not practical everywhere on the landscape and so are not for all homeowners seeking an alternative energy supply. One way this geographical limitation of wind power could be lessened would be the widespread adoption of what is being called "the reverse meter" principal, wherein a homeowner could put power from the wind back into the public system and thereby reduce his bill. The power companies would in effect be buying electricity from the individual producer, who would still be able to use as much electricity as needed.

Yet even without economic incentives of this sort, the grassroots wind power movement is growing, perhaps faster than the machines involved. Domestic commercial production of wind generators is still at a very early stage. As yet there is only one complete unit made in this country, the "Windcharger" by Dyna Technology, rated at one-tenth the output of the Dunlite. Other systems available here come from Switzerland and France. The Swiss "Electro" wind generators range in output from .8–6 kw (one kilowatt equals 1,000 watts) and cost up to $6,000. These machines are sold in the West by the Real Gas & Electric Company of Guerneville, California. Two years ago, Solomon Kagen purchased a Dunlite for his ranch and became so taken by the idea of wind power, that he founded RG&E to make complete wind generator systems and their installation more easily available to the public.

Another firm in Texas distributing a French version, the "Aerowatt" in models rated up to 4 kw. They are described by Automatic Power of Houston as "highly reliable and efficient, but expensive." Four kilowatts of Aerowatt run over $20,000.

What is sorely needed is the beginning of production in this country to lower the initial cost and to make available an array of designs to meet varying environmental and energy needs. This seems certain to happen as the cost/output ratios of other energy systems continue to rise, making the wind a more economical and feasible energy source.

The surge of interest in owning wind generators has been accompanied by an encouraging increase in research and development. Coming out are kits and all manner of do-it-yourself plans; some aim at economy, some for greater efficiency. Prototypes are appearing everywhere in industry, universities, and among independent inventors.

High up on Sonoma Mountain Road near Santa Rosa sits a wind generator atop a short tower made of 4×4's. Smaller than the Taylor's Dunlite, this handmade unit is the pride and joy of self-taught wind power enthusiast Jerry Forcier. Typical of the growing fraternity of backyard innovators in this exciting field, Jerry has devoted much of the past 12 months to studying wind mechanics and putting together a unit which incorporates what he believes to be significant improvements over existing equipment.

"Inefficient!" is what Jerry says of many of the design characteristics of currently available wind generators. With his eye on thrift and economy of operation, efficiency is the key. The demand for it goes from first catching the wind to the nature of the final product. His propeller is like others in number of blades (3) and in its automatic feathering device to protect the generator from too-high rpm's in heavy winds. Jerry's blades are custom designed, however, to meet the lighter-than-usual local wind conditions. He's made them almost twice as wide, about 14 inches, as common blades for better wind catching in light airs.

A major departure of his model is the direct-drive feature, eliminating the common step-up gear box. "It takes a lot of the wind's energy just to run gears; maybe as much as one-fourth of the energy caught by the prop is eaten up by gears. True, they make the generator turn faster for more output, but greatly reduce its effective life. With direct drive the slower-turning generator's life is extended three to fivefold." There is a drop in output at lower rpm's, but Jerry feels he has increased output efficiency of this unit with special rewinding of the rotor (which turns inside the stator). By adding many pounds of copper wire to the rotor, he has partially offset the output drop of his lower rpm generator. Avoiding the gears also means less maintenance, fewer trips up the tower.

"I think I've got about a 3–3.5 kw generator here," says the inventor—certainly enough to power the household's lights, appliances (minus cooking and heating), and tools, given adequate wind.

"The electricity from it will be used in the form of direct current, without inverting to a.c. Producing d.c. just takes a lot less raw energy than is needed for a.c. production." We all live with a.c., but except for TV's and stereos and a few other unadaptable items, d.c. power could do it all.

Jerry Forcier's wind-work and that of others across the country is answering a growing demand for decentralization and self-sufficiency in power production. People are fast becoming more energy-conscious as the limits of mass-produced energy from nonrenewable natural resources become apparent. Three years ago, there was only one source of domestic-sized wind plants in the country. Today there are about a dozen, and more are certain to follow soon.

In Pursuit of the Zero-Discharge Household

•

Steve Smyser

Except for residents of the arid West and Southwest, Americans have always taken clean water for granted. Each generation has found new ways to use water faster, until the average daily use per person has risen to approximately 65 gallons. At a time when 70 percent of the human race has no piped water at all, the annual water consumption of the average North American family is 88,000 gallons. With new households hooking on to our already overtaxed sewer mains at a rate of between two and three million a year, Americans will need 800 billion gallons of water a day for domestic consumption by 1980—twice the amount used in 1974. Municipal sewage loads are expected to quadruple in the next 50 years.

When nineteenth-century public health officials were forced to come to terms with the environmental problems caused by the flush toilet, they opted for huge central sewage systems, unable to conceive of a day when it would take tens of billions of dollars just to avoid losing ground more rapidly in the area of sewage treatment and pollution control.

Today, there are many who share California architect Sim Van der Ryn's view of current waste-management practices as a prescription calling for "one part excreta mixed with 100 parts clean water. Send the mixture through pipes to a central station where billions are spent in futile attempts to separate the two. Then dump the effluent, now poisoned with chemicals but still rich in nutrients, into the nearest body of water. The nutrients feed algae who soon use up all the oxygen in the water, eventually destroying all aquatic life that may have survived the chemical residues."

Tragically, this same excess of sewage that is fouling our rivers and oceans is sorely needed to restore fertility to depleted soils—a precious

HOW WATER IS USED BY A TYPICAL FAMILY OF FOUR

Use	Gallons Used per Day
1. Dishwashing	15
2. Cooking, drinking	12
3. Utility sink (washing hands, etc.)	5
4. Laundry	35
5. Bathing	80
6. Bathroom sink	8
7. Toilet	100
Total	255

resource being permitted to charade as a "waste." Current wastewater technology is antagonistic to the equilibrium that must exist between nutrients being removed from the land and being returned, between the forces of decay and the forces of renewal.

The chief villain in this depressing scenario, of course, is that paragon of civilized society, the flush toilet. About 41 percent of all water piped into homes is used to flush toilets. At an average of slightly over five gallons per flush, that means the typical user of a flush toilet contaminates 13,000 gallons of pure water a year to carry away 165 gallons of body waste.

To the family that routinely composts its kitchen wastes, recycles its paper, glass, and cans, keeps the thermostat in the 60s, and generally takes pride in maintaining a no-waste organic household, the inconsistency of having to flush away all that clean water can be downright painful—a glaring gap in an otherwise closed loop.

Now that the true energy and environmental costs of overloaded sewage treatment plants and of securing fresh water are beginning to be understood, what are our options at the individual household level? Clearly, it's going to be more economical to cut water consumption than to develop new sources, but just how much of an impact can we really hope to have by adopting water-conserving habits and hardware around the home?

One of the best known and most effective methods of saving water in flush toilets is to place an ordinary brick in the toilet reservoir, a simple act that displaces a little over one quart of water and saves that much every time the toilet is flushed. Two 1-quart plastic bottles, filled with water and weighted with a small stone or some other object to keep them in place, accomplish the same thing. When the toilet is flushed, the bottles simply hold back the volume of water they contain. Unlike bricks, the plastic bottles won't break the toilet reservoir if accidentally dropped.

There is also available a wide array of commercial toilet reservoir-volume reducers and improved float assemblies. The volume reducers, which work by walling off a section of the toilet reservoir, are inexpensive and easily installed. The new float assemblies can be adjusted to maintain a lower water level in the toilet reservoir without any reduction in flushing efficiency. A recent study in the suburban Washington, D.C., area indicated that the use of these wasteflow reduction devices by themselves produced a savings of up to 26 percent in household water use.

If you're remodeling or building a new bathroom, you can put a real dent in your wastewater output by installing one of the new shallow-trap toilets. By virtue of its smaller water reservoir and the special design of its bowl, the shallow-trap toilet uses about 3½ gallons per flush as compared to five or six gallons for the standard toilet—a savings of about 30 gallons a day for a family of four. A growing number of local sanitary districts are following the lead of the Washington, D.C., Suburban Sanitary Commission in requiring the use of 3½-gallon flush toilets in new housing and other construction.

Using five gallons of clean water to dispose of ½-pound of feces is extravagant enough. Using (and paying for) the same amount to flush away one pint of urine is patently absurd. There are now several models of dual-flush toilet devices on the market that reduce the flush cycle to 2½ gallons for solids and 1¼ gallons for liquids. The different cycles are initiated by a short, sharp pull on the flush handle for the smaller amounts of water, and a longer, more persistent pull for the larger amount of water. Although similar in effect, the several dual-flush devices on the market differ substantially in design. Some are easily incorporated inside the conventional toilet tank; others require modification.

After the toilet, the heaviest water user in the household is the shower. Approximately 30 percent of the total household water consumption goes for bathing—roughly 80 gallons a day for a family of four. This amount can easily be cut in half by the installation of any one of a variety of devices that limit the rate of flow from shower heads and faucets. Whereas normal flow from conventional shower heads ranges from 5 to 15 gallons per minute, flow control devices cut this rate to 3½ gallons per minute with no sacrifice in cleansing or general acceptability.

For areas in which water is in extremely short supply, there are rather expensive units that reduce shower flow to ½-gallon per minute by mixing the water with compressed air. Depending on water pressure and the length of your showers, a flow-restricting device could save your family 2,000 gallons a year per person. There's no real need for the pressure in the household water line to be 140 pounds per square inch—60 psi works just as well.

Besides the water savings, a reduced-flow shower can save you plenty (some studies indicate as much as 21 percent) in hot water heating costs. In the typical household energy budget, the energy required for domestic hot water heating ranks second only to that used in heating the home itself.

Faucet aerators and spray taps for the sink are already in fairly wide use. Unlike the conventional faucet, which allows the water to gush out in a single stream, these devices mix air with the water as it leaves the faucet, which gives the illusion of more water flowing from the tap than actually is. Faucet aerators are inexpensive, easy to install, and save the family of four about two gallons a day.

As anyone who has been scorched or frozen while trying to adjust the shower temperature is well aware, considerable water goes down the drain before the bather can even enter the shower fully. A product that eliminates this waste is the thermostatic mixing valve. This valve mixes hot and cold so that it issues from the tap at a preset temperature, a virtue that also lets you turn off the water while soaping without having to readjust the handles.

Automatic clothes-washing machines account for about 15 percent of the water consumed in households where they are present. Front-loading models use 22 to 33 gallons per cycle; top-loading machines required 35 to 50 gallons. Washers with suds savers use less water by providing for the reuse of wash water for a second load. Some automatic washers allow the amount of water used to be adjusted for load size.

Automatic dishwashers use water lavishly, too—from 13 to 19 gallons a day. The best conservation technique is simply to load to capacity for each use.

Of the various water-consuming devices in the household, the dishwasher is the only one that really justifies the 140-degree F. water that Americans maintain in their water heaters 24 hours a day. A recent Stanford Research Institute study concluded that a whopping three percent of all the energy consumed in the United States goes for heating domestic hot-water tanks. Hopefully, the hardware will soon be available in this country to enable us to follow the lead of West Germany and France in installing small "point of use and heat on demand" water heaters above faucets in our kitchens, bathroom sinks, and bathtubs, where hot water must presently be mixed with cold to avoid scalding.

To achieve the maximum water savings from these devices, they should be used in combination. William Sharpe, Water Resource Specialist at Penn State, calculated the savings that could be expected from the combined use of a volume reducer in the toilet, a flow-control device in the shower, and aerators on two sinks. In the case of a family of four, he found that a total of 76 gallons of water could be saved per day, amounting to an annual savings of $47.57 on water, power, and sewage costs.

In trying to calculate how much of an impact the widespread adopting of water-saving devices might have, it's important to remember that about one of every three homes in the United States still lies beyond the sewer mains. When it comes to conserving water, the homeowner with a private sewage system has a definite advantage over the urbanite forced to pay daily tribute to the communal sewer. He is the real beneficiary of the great strides being made in decentralized sewage disposal and toilet technology.

There's an experiment underway in Boyd County, Kentucky, that could produce some important answers in the quest for ecological home disposal systems. The central attraction in the Kentucky demonstration is a recently developed aerobic alternative to the conventional anaerobic (oxygenless) septic tank. To show how non-polluting technology can improve home sanitation, 36 of the new units are being installed in homes along the banks of the Upper Chadwick Creek under the auspices of The Appalachian Regional Commission, a federal-state agency charged with improving the quality of life in 13 eastern states.

In the aerobic units being demonstrated, a small pump mixes air into the wastewater, thereby breaking up solids and promoting the growth of air-loving bacteria that further digest the wastes. What's involved, in effect, is a backyard version of a secondary biological treatment plant. The effluent leaving the aerobic tank is similar in quality to that resulting from secondary treatment, and filtration through a few inches of soil in the drainfield is equivalent to tertiary treatment. Unlike the typical septic tank, the aerobic tank does not have to be pumped out every year or two, and its greatly reduced drain field is not prone to clogging.

Upon leaving the tank, the cleansed water can be dissipated in a number of ways: a) through a tile-drain field under the lawn (as with a septic tank); b) through a sand filter and then into a stream or drainage ditch; c) into an evapo-transpiration bed; d) to an above-ground spray field; or e) directly back to the house for reuse in the toilet tanks. Each of these methods will be tested in the ARC project, and water quality changes in the Upper Chadwick Creek will be monitored closely.

Besides the single household units, two larger aerobic systems will be installed—one serving five houses and one serving six. Initial costs for the 36 units (to be supplied and installed by six different manufacturers) will be borne by The Appalachian Regional Commission, and the local sanitary district will handle the periodic inspection and maintenance.

Aerobic treatment units of the types being used in the ARC project are presently being produced by a dozen major manufacturers in this country. Installation costs range from $1,000 to $3,000 a house. Two companies Multi-Flow, Inc. of Dayton, Ohio, and Thetford Corp. (Cycle-Let) of Ann Arbor, Michigan, are marketing aerobic systems that filter the effluent to the point where it can be safely cycled back to the flush toilets.

Thanks largely to the search for improved sewage-handling systems for seasonal dwellings and homes built in areas with unfavorable soil conditions, there is emerging an impressive range of private waste-handling systems that go the aerobic units one better—by eliminating water altogether.

Probably the best known of these designs is the composting toilet, introduced to the United States by the Swedish inventor Rikard Lindstrom, and presently being distributed by at least three different companies here. Composting toilets operate aerobically to reduce wastes biologically to an inert compost which is periodically removed and returned to the soil. Models range from large fiberglass devices, designed to handle both human wastes and kitchen garbage, to small, self-contained devices that require electrical heat and ventilating systems to maintain the composting process.

Incinerating toilets are designed to burn up the solid waste and evaporate the liquids, leaving behind sterile ashes. Most models run on either electricity or propane gas, and have blowers to remove odors, heat, and vapors. Incinerator toilets are competitive in price to conventional septic-tank systems, but power and fuel costs largely rule out their everyday use.

At least one manufacturer in this country now makes a recycling toilet based on biological principles of waste digestion: body wastes and toilet tissues are turned into water by the action of enzymes and bacteria. According to the company, the resulting effluent can be easily disposed of in a small drain field or dry well near the house. Maintenance is limited to a weekly addition of a package of freeze-dried enzymes and bacteria, and a renewal of the charcoal filters once every two years.

Another waterless toilet design that is being viewed with increasing interest for single-family installations is the oil recirculating system, in which wastes are carried by a low-viscosity mineral oil to a gravity separation tank. In the tank, the oil floats to the top and the water-saturated wastes sink to the bottom. The oil is drawn off from the top through a series of filters and then recirculated to the toilet. Wastes are periodically drained from the bottom of the tank and disposed of either by incineration or aerobic digestion and drying. In a typical household, a 400-gallon holding tank can store enough oil to last a family of four for a year.

Obviously, no one waste-disposal or recycling system offers a solution for every situation; what's needed is a combination of approaches. Hopefully, these recent innovations in water-saving and waste-handling technology will coincide with a heightened public concern for wise water utilization.

The Clivus Multrum Composting Flushless Toilet

•

Ray Wolf

When news of a Swedish composting toilet began to circulate in this country a few years ago, it sounded like the perfect answer to the problem of dealing with human waste—no moving parts, no water or energy used, no chemicals, no liquid effluent, no plumbing, no cleaning, and no smell, plus the fact that the unit produced a rich, humus-like fertilizer. But unfortunately, the Clivus Multrum has not been an overnight success, selling only about 250 units in the first few years of business here.

"Most people continue to view gravity-flush toilets as outhouses, which they aren't," says Bob Kaldenbach, marketing director for the U.S. Multrum Company.

Far from being an outhouse, the multrum is a modern, efficient method of treating human waste and organic kitchen scraps through biological action. The unit is entirely self-sufficient.

The Multrum utilizes an impervious fiberglass container to treat waste. Directly above this tank is a toilet seat, and a small distance away, a garbage chute. A vent pipe is located between the two incoming chutes. The main tank is broken into three chambers at a downhill angle, one receives toilet waste, the other kitchen waste and the third is a storage chamber for the combined compost. The unit operates by letting toilet waste slowly slide downhill into the kitchen waste, the two combine and further compost and slide into the storage area.

The vent pipe must be kept warm to prevent water vapor from condensing and running back into the heap. The unit cannot handle large amounts of water.

One drawback of the Clivus Multrum is that separate facilities must be provided to handle dish, laundry, and bath water. However, the unit

113

can be adapted to handle added water, but additional dry matter must be added to compensate.

Usual installation of the unit is in a basement, but it may be placed on the first or second floor of a home, in a partial basement or outside as in a comfort station for campgrounds, or under an outhouse. However, precautions must be taken to prevent the unit from freezing during winter months.

With the unit in the basement of a home, the toilet pedestal would be located directly above the top end of a unit. The kitchen waste tube must be placed on the opposite side of the toilet wall, allowing the waste chute to be above the second chamber of the composting tank. This should not be inconvenient in most cases, as houses are designed with kitchens and bathrooms sharing walls, to reduce plumbing costs.

Before being put into operation, the Clivus must be primed. This involves spreading a layer of peat moss four to five inches thick over the entire bottom of the unit, followed by two inches of garden soil and two inches of cut grass, dead leaves, or garden garbage. The peat soaks up urine and its ammonia, and the filtering process of soil and garbage by the peat favors de-nitrifying bacteria leading to an increase in the range of species specializing in the breakdown of organic matter.

By the time the urine reaches the garbage heap, it is transformed into nitrate, which can be readily used to break down the cellulose which dominates the composition of most household garbage.

As garbage and effluent combine and break down, it slowly slides downgrade into the storage area. By the time the composted material reaches the holding area, it is reduced to some 10 percent of the original volume. In Sweden, where more than 1,000 units have been in operation, some for over 10 years, tests show that for each person using the unit full-time, it will produce 60 pounds of fertilizer a year. An analysis made in Finland of compost from a Swedish Multrum showed an NPK content of 20-12-14.

About this far along, anyone who has ever used a privy will be thinking, "That's all well and good, but the thing must really stink." When properly installed, a Multrum will not smell, thanks to a very good air-circulation system. A natural downdraft is created whenever the toilet or garbage chutes are opened, pulling room air down through the system, preventing any aroma from going up and into the room. A venting system within the tank constantly pulls air through the composting waste, to ensure that the composting is aerobic and thus free of offensive odors.

While the composting is going on in the chamber, the unit is practically maintenance-free. The garbage and excrement chambers never need be emptied, and the finished compost storage area can be emptied only as you need compost. It takes from two to four years from when the unit goes into operation before any material is ready to be removed. The storage area can hold a ten-year supply of compost for a family of from four to six without being emptied.

The long process that waste goes through is responsible for the safety of the product. Unlike the typical compost heap, temperatures in the chamber seldom go above 90 degrees F. Bacteria in the chamber operate best at about body temperature, which is not hot enough to destroy most pathogenic organisms. What does kill these organisms is the amount of time they are treated. The two- to four-year detention period subjects pathogens to predation from other organisms within the chamber, especially the biological life that destroys the cellulose content of the garbage, in addition to a climatically unfavorable (for the pathogens) environment created within the unit.

Presently, the Multrum is usually limited by its design to installation in custom-built homes. This stems from the need to have the bathroom directly above the tank and the kitchen garbage chute within a few feet of the toilet. This also restricts the unit to only one toilet, unless the toilets are stacked directly above one another in a two-story home.

Carl Lindstrom, son of the Multrum's inventor, recently designed a horizontal transport system that is in use in Sweden, and has overcome all these problems. The horizontal transporter consists of a four-inch plastic pipe with an electrically driven, hollow screw inside. The horizontal transporter requires about ½ cup of water per flush, and the motor runs for about 30 seconds per flush.

Although this will solve most of the unit's current design problems, in the eyes of a purist, it no longer is an ecologically perfect unit, as it requires both electricity and water to function. However, the unit does not have to be used with the horizontal transporter, which is sold only as an option to help overcome structural installation problems.

The real problem with the Multrum is not in its design, but with getting public health officials to approve installation of the unit. Maine approved installation of the units providing an additional system is available to treat bath and dishwater from the household. The Multrum does not treat this type of waste, only organic waste. The Maine statute allows for a 40 percent reduction in the size of a septic tank's leaching field if a Multrum is used.

Massachusetts law prohibits use of a septic tank to treat bath and dish water excluding effluent in the tank, which precludes use of a Multrum. In other states, health officials have expressed concern over the safety of the compost, the gases emitted from the vent pipe, and the safety of having such a unit in the household. "We are pushing against some deeply ingrained prejudices by public health officials who view any alternative to conventional flush toilets as at best suspicious, at worst subversive. It will be a struggle, but it's one worth making," Kaldenbach said.

Of the 35 units sold and installed at this time, most have been put in houses designed to use such a unit. They have either been approved by a local official, or get around the health codes by also providing a standard toilet. "Most codes require you to have one approved toilet facility, and they don't get involved with any additional units," Kaldenbach explained. Multrums are now installed in 16 states.

Kaldenbach is tackling the problem of getting approval of the units on a state-by-state basis. This will be done simultaneously with the establishment of a nationwide distributing system. Prospective distributors need not put up any cash investments, for a nonexclusive agreement, on a commission basis only.

New distributors are asked to buy and install a Multrum, but this is not a requirement. The distributors will work with state and local health officials for approval of the unit. Research is being conducted in Sweden on Multrums to prove the safety of the compost, and their efficient operation in the home. The company has contracted with Barry Commoner's Center for the Biology of Natural Systems to do research on the units. In one case, the waste will be inoculated with known pathogens, and will be tested to see if the process can kill off the pathogens in the waste. When the results of these tests are in, Kaldenbach feels certain the units will be approved.

The addition of the horizontal transporter and inclusion of a water flush, regardless of how small, should also make the unit more acceptable to health officials in Kaldenbach's estimation.

Without the horizontal flush, the Multrum will save the average family of four from 40,000 to 50,000 gallons of fresh water a year. This is about half of a family's yearly use of water. The small amount of water used for the horizontal flush should not amount to much, as it is only ½ cup, compared to the conventional toilet's five-gallon flush.

For the organic gardener, the Multrum offers a good method of on-site waste treatment. Perhaps the only drawback is that the unit may be a bit too efficient, in that the amount of fertilizer returned from the unit would not go very far on a productive garden. Most of the nutrients are volatilized in the chamber and go up the exhaust stack in the form of harmless water vapor and carbon dioxide.

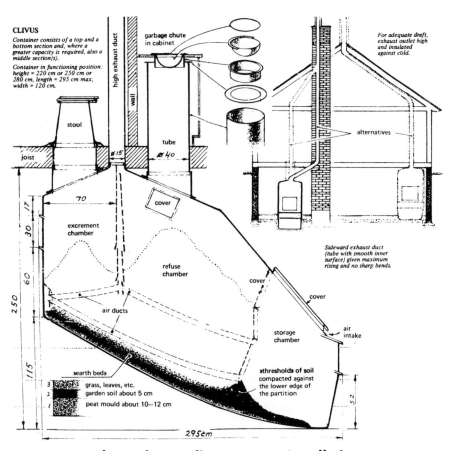

CLIVUS

Container consists of a top and a bottom section and, where a greater capacity is required, also a middle section(s).

Container in functioning position: height = 220 cm or 250 cm or 280 cm, length = 295 cm max, width = 120 cm.

high exhaust duct

garbage chute in cabinet

For adequate draft, exhaust outlet high and insulated against cold.

wall

stool

joist

tube

alternatives

ø 15   ø 40

70   cover

excrement chamber

refuse chamber

Sideward exhaust duct (tube with smooth inner surface) given maximum rising and no sharp bends.

cover

cover

17
30
60
250   air ducts

storage chamber

air intake

5.2

115

»earth bed»

thresholds of soil compacted against the lower edge of the partition

3 grass, leaves, etc.
2 garden soil about 5 cm
1 peat mould about 10–12 cm

295 cm

The Multrum Clivus—Home Installation

Recycling the Wash Water

•

Frances Kilbourne

Doing the laundry for a family of four, two of whom are in diapers, uses a lot of water. But we have developed a plan that not only cuts down on our water bill, but also benefits our garden and helps, in a small way, to conserve some purified water.

The equipment we use consists of a compressor from an old refrigerator, a discarded 22-gallon water tank from a pressure system, and the garden hose.

After the clothes are laundered with a less-than-one-percent phosphate soap like "Ivory Snow," I drain the wash and rinse water into the tank. (I use a wringer-type washer because it does the entire job with one tub of water compared to several used by my automatic.)

The compressor is permanently connected to the water tank, along with a length of garden hose reaching out the basement window. Plugging in the compressor gives me 22 gallons of "free" water for the garden.

My vegetables and flowers have thrived with the extra watering while the mild, soapy mixture helps keep down insect and fungus infestations. (Remember grandmother's prize roses that never saw a modern chemical, but got the dishwater thrown on them three times a day? It's the same principle.)

Now we are working on a plan to divert our kitchen sink and bathtub drains into the water tank and use that water, too. We'll have the cleanest garden in town.

Modern Living Without Garbage Service

•

Wayne H. Davis

"I canceled the garbage collection," my wife said one day in June 1971. "Asked the men to take the smelly old cans with them and said we just don't want to be bothered with garbage and trash anymore."

Although we had discussed the possibility of doing without garbage collection, the reality of the action was rather frightening to me. Raised in the city, I was thoroughly addicted to garbage cans. Our household regularly filled two large cans in anticipation of the twice-weekly collection, and there was scarcely a day that I didn't carry out one or more large grocery bags full of garbage. The stinking cans, the swarms of flies seeking putrifying food scraps on hot summer days, and night rattling of cans by raiding dogs was accepted as part of the way of life in a modern society. The thought of getting away from it all was pleasing, but I was afraid that after a short effort at handling our own garbage we would sheepishly request service to resume.

Now with a few years' experience behind us we can pronounce this venture a success. For those who would consider trying it, I shall explain why we did it and how our system works.

One consideration was money. The several dollars per month we paid for garbage and trash collection may seem a small part of a budget, but it is considerably more than we pay for our daily newspaper. Stopping it was a logical step in our savings efforts to reduce nonessential services, such as professional haircuts, and hiring a plumber to fix a leaky faucet.

The most important factor, however, was a desire to find an alternative to what I consider a self-destructive life-style. This life-style grew

Dr. Wayne H. Davis is a Professor of Zoology at the University of Kentucky.

out of the country's "use it once and throw it away" economic philosophy which replaced the hated rationings of World War II days and led economists into the practice of measuring progress by the rate at which we turn our natural resources into junk. As an ecologist I recognized this as false doctrine. No living species can survive long without recycling from one generation to the next the minerals it needs. Man's system of moving nutrients from the farm to the ocean by way of the sewers, and of moving industrial materials from the mines into landfills is not a viable life-style.

When we began our experiment only newspapers were easily recycled here in Lexington, Kentucky. However, since local activists had promised a recycling center for glass within six months, jars were relegated to our basement for this duration.

We cut back on purchases of food in cans and jars, going heavily to the more wholesome and often cheaper fresh produce. We had already whipped the problem of the soft drink can and bottle, which seems to make up a major portion of the average American's shopping basket. We never buy soft drinks, using homemade root beer instead. With root beer extract, which can still be purchased at a very few of what I consider our more progressive groceries, and a collection of soda bottles with screw-on caps, I can make 18 quarts in 20 minutes at a cost of about a dime per quart.

At first I burned our few cans on the grate and put them in the garden to rust. I quit this after reading in a British journal about the poisoning of a truck garden after 30 years of dressing with municipal compost. The plants were poisoned by lead, zinc, and cadmium from shredded cans in the compost.

For a while we buried cans in the yard, but were most relieved when our recycling center began taking and putting them back to work.

Today our system works like this: Newspapers, glass, aluminum, cans, and pasteboard go to our recycling center, which is open just one day each month. Magazines go to hospital waiting rooms. Kitchen garbage is buried in the garden. Paper and plastic litter is burned on the grate and the ashes put in the flower beds. (I used to put ashes in the garden, but quit upon learning that colored printer's inks contain rather large amounts of lead.) Our occasional large pieces of scrap we take to a scrap dealer.

Here is a rundown of our procedure: In the kitchen we now have two garbage sacks. One holds both glass and cans which we separate at the recycling center. We rinse, remove labels from, and flatten the cans. Our second sack is fitted inside a plastic garbage basket with a folded newspaper beneath it. Into it goes all kitchen scraps. When it fills I take it to the garden and bury the contents, bag, newspaper, and all.

Since both bags fill slowly, little work is involved in handling them. Cans and glass are carried out about every two or three weeks and the kitchen scraps about weekly. If the garbage begins to stink, my wife simiply closes the top of the bag. This contains the odor and is a signal for me to remove the bag.

The least satisfactory part of our system deals with paper and plastic scrap. We have been unable to reduce the volume of such packaging and junk mail that comes into our house, and must burn it in our basement fireplace about twice a week. This creates some air pollution, which would not be tolerated in many communities. We hope that recycling of plastic and paper scraps will someday become feasible and end our trash burning.

Putting garbage into the garden has never caused serious problems. Directions for composting say not to use grease or meat scraps because they attract dogs and rats. With my unfenced yard I found dogs would dig up the garbage. At first I countered this problem by covering each site with stones. Then I discovered that a single old tire is satisfactory. After a burial site has been covered for a week it is no longer attractive to dogs, so I just move the tire when I bury the next bag.

Rats are a problem in my yard and garden. I must eliminate them each time they get established, which may be two or three times a year. However, the garbage does not seem to be a factor; only once have they ever dug into it. The winter bird-feeder and the lush cover and food of the late-summer garden seem to be the main attraction for rats.

When the soil is frozen, I bury the garbage under a pile of leaves where the ground is always soft. In the hot summer, when our garbage is entirely vegetable matter unattractive to dogs, I simply throw it between the corn rows and cover it with the grass clippings my neighbors put out for the trash collector to pick up.

Buried garbage decomposes quickly. It is attractive to earthworms and other invertebrate soil conditioners. By three to four weeks nothing can be found of the garbage except scraps of eggshells and chicken bones—and even these disappear with more time. My garbage has been a welcome addition to the leaves and manure which I have used to turn a plot of clay from my basement excavation into a fine garden with excellent soil texture and fertility.

There are disadvantages to our system. Preparing cans and jars for recycling is extra work. The extra bags to separate garbage, trash, cans, and jars are a nuisance. Some of our garbage space is needed to store materials between trips to the recycling center. However, we consider our effort a success, and think that others interested in organic living might want to try it. I doubt if we will ever go back to garbage cans and collection service.

Of this country's Total National Trash Can (125 million tons of solid waste a year), food wastes account for some 17.6 percent (22 million tons). Of this amount, 17.8 million tons, or 72 percent, come from private households. That's 593 pounds of food waste per household per year—the energy equivalent of 26.1 gallons of gasoline.

As petroleum inputs drive fertilizer prices even higher, the mentality that condones burning, burying, or flushing this wealth of nutrients into our seas, while at the same time shelling out $6 or $7 a gallon for synthetic fertilizer, becomes as indefensible ethically as it is economically. Twenty-two million tons of wasted food a year could feed a lot of undernourished people in this country and abroad.

The potential of food waste as organic fertilizer is demonstrated by its content of the three essential elements, nitrogen, phosphorus, and potash.

	Nitrogen	Phosphoric Acid	Potash
Eggshells	1.19	0.38	0.14
Coffee grounds	2.08	0.32	0.28
Orange skins	—	2.90	27.00
Grapefruit skins	—	3.58	30.60
Apple cores	.05	.02	.10
Chicken bones	—	—	—
Banana skins	—	3.25	41.75
Lettuce	—	—	—
Cantaloupe rind	—	9.77	12.21
Corncobs	—	—	50.00
Carrot tops	0.40	0.40	3.00

Overall, fresh garbage usually contains an analysis in the neighborhood of 2.0 to 2.9 N —1.13 to 1.30 P —0.8 to 2.2 K.

Heating Your Home
Without Harming Nature

•

Jeff Cox

There's a question that must be asked by anyone who'd like to use wood from a woodlot on a sustained yield basis to heat his house.

The question is: How many acres of woods will yield enough firewood to heat the house for a season, so that the amount taken is equal to the amount that grows back each year?

The answer can be figured from a formula which will be presented shortly. But first, some background on heating with wood.

For many people on homesteads or in the country, wood heating offers a great environmental advantage over coal, gas, oil, electricity, or power from a nuclear plant. Coal must be dug from the earth, an irreplaceable natural resource, and the digging often irreparably scars the land. Oil, too, is a dwindling resource, as attempts to open up oil fields in the fragile arctic show. Gas is similarly limited. Electricity is usually made from dynamos powered by coal or oil. Power made from nuclear plants is probably the worst of all.

It is possible to manage a woodlot to avoid all these methods of heating. A large enough woodlot will supply you with enough wood to heat your house, and nature will grow back each year what you've taken. But how big does the woodlot have to be?

It all depends, of course, on many variables—how hot you keep your house, how big the house is, how well insulated it is, how cold the winter is, how many British thermal units (Btu's, a measure of heat energy) are contained in the wood you're burning, the efficiency of your heaters, etc.

Jeff Cox is the Executive Editor of Environmental Action Bulletin, *and an Associate Editor of* Organic Gardening and Farming *magazine.*

All these variables can be simplified and reduced to the following equation:

$$\frac{XYZ}{M} = C$$

C is the number of cords of wood needed to heat your house for a season. You can figure the size of the woodlot needed on this basis: one acre of average-growth hardwood yields one cord of wood per year on a sustained yield basis. That is, one acre grows about a cord of new wood each year. A cord of wood is a stack 4×4×8 feet, or 128 cubic feet.

We're going to assume that you like to keep your house at about 70 degrees F., maybe a shade cooler. The equation also assumes you have wood heaters of 50 percent efficiency, which is pretty high. If you plan to burn wood to heat your home, either an Ashley or a similarly efficient heater is recommended. Fireplaces run at about eight percent efficiency, and many old stoves run at only 15 percent. To avoid unmanageable amounts of wood, get good, efficient heaters that send at least half of the heat produced into the house instead of up the chimney.

In the equation, the letters have the following meanings:

X is the number of square feet of floor space in your house, not including unheated portions such as attic or cellar. The average small house contains about 1,200 square feet. You must determine the figure for your house by measuring.

Y is the number of Btu's needed to heat one square foot per hour. For a well-insulated house, this figure is 28 Btu's. For an uninsulated house, 45 Btu's. Good insulation will pay for itself many times over in a short time, so if you plan to heat with wood, good insulation is a must. If your insulation is so-so, choose the figure between 28 and 45 that you think reflects the quality of your insulation. A merely adequately insulated house might call for a 34 Btu figure.

Z is the number of degree days per season in your area. Degree days are the average number of degrees of temperature below 70 degrees F. on a given day. For instance, if the average temperature on January 15 is 38, you subtract 38 from 70 to get the degree days—32 degree days in this instance. But don't get panicky. A call to a gas, oil, or electric company, or the local weather bureau, can get you the degree-day figure for a heating season for your area. For example, in our part of eastern Pennsylvania, the number of degree days for a normal season is figured at 5,800.

M is the number of Btu's contained in a cord of the wood you're burning. Consult the chart accompanying this article for the Btu content of various woods. In a mixed stand, take an average based on the composition of your pile of wood. If it's half red oak and half ash, then the cord would contain 21,150,000 Btu's.

Now let's set up a hypothetical homestead to see how the equation works. Let's say a man in eastern Pennsylvania wants to use wood to heat his home, which has 1,200 square feet of floor space. It's also well insulated, so we'll use the figure of 28 Btu's per square foot per hour. The number of degree days in his area is 5,800 per season. Let's say he's using a mixture of white oak and red oak, which rates at 22 million Btu's per cord. Plugging all these figures into our equation, we get:

$$\frac{1,200 \times 28 \times 5,800}{22,000,000} = 8.85$$

This works out to about nine cords of wood needed to heat for a season. Since one acre yields a cord on a sustained yield basis, you'd need a nine-acre woodlot at least. Of course, if your woodlot is on a dry slope, the trees may grow slower. Your state forester, whose headquarters is located at your state capital, can give you literature on planting and managing a woodlot, as well as evaluate the growth rate of your stand. Your county agricultural agent will also have material on the management of woodlots, both for producing firewood and for optimum wood production for sale.

BTU CONTENT OF ONE CORD OF VARIOUS WOODS

Kind of Wood	Btu's per Cord
Hickory	24,600,000
White Oak	22,700,000
Beech	21,800,000
Sugar Maple	21,300,000
Red Oak	21,300,000
Birch	21,300,000
Ash	20,000,000
Red Maple	18,600,000
Elm	17,200,000
Yellow Pine	18,500,000
White Pine	13,300,000
Aspen	12,500,000

BURNING QUALITIES OF COMMON WOODS

Sam Ogden, the Vermont woodsman, reminds all who would heat with wood that seasoned wood is best. Softwood and rotten wood, as well as green wood, don't burn as well. Give a woodpile of freshly split green wood about four to six full months to dry out under cover.

Here are Sam's recommendations:

Elm — Not desirable; the wood's too soft and is hard to split.

Apple — One of the best; it burns clean and hot, and gives off a delightful fragrance.

Pear — Similar to apple.

White birch—Okay, but its oily bark burns with explosive violence.
Yellow or Black Birch — A superior fuel.
Hard maple — Good all-round burning characteristics.
Red or soft maple — Not as Btu-packed as its harder brethren.
Black cherry — Burns hot but it pops out sparks.
Ash — Same as black cherry.
Beech — One of the very best firewoods.
Oak — If you're burning white oak, you're burning the best.
Willow — Avoid this softwood. Btu content is low, erupts and sparks as it burns. This holds for other softwoods such as pine, poplar, hemlock, etc.

WOOD SCAVENGING CAN HELP FILL THE WOODBOX

One of the nicest things about wood fuel is that it is a renewable resource. Properly managed woodlots will supply your fuel needs with no damage to the environment—esthetic or otherwise. But before you start cutting growing trees, or even mature trees, you might take a clue from one Pennsylvania Dutch farmer who gets all his fuel free and seldom fells a tree. Behind his house is a woodpile bigger than some houses. Where does it come from? Well, whenever the telephone company clears the right-of-way, he's there with his chain saw and wagon to clean up the residue. If a storm blows a tree across the road, he gets it. If a neighbor wants a tree removed, he'll oblige. When a new building goes up and trees are sacrificed, he sees that they aren't wasted. Even if the opportunity comes in the summer, he doesn't let it pass. Consequently his woodpile is always well-stocked. There is something particularly warming about wood that was destined to be wasted.

How to Build a Fireplace Fire

•

Raymond W. Dyer

The secrets of all fires, either fireplace or stove, are first having good, hard, dry wood; and second, giving the fire as little draft as possible. It is the careful control of the draft that maintains a good fire of glowing wood coals. When the fire is started with paper and kindling, both drafts should be almost open.

Properly built, a fireplace warms the heart and body in equal measure. Most fireplaces are greatly misused by burning fires too blazing and hot. The mistake begins with the thinking that the andirons are a grate upon which the fire should sit, so it has a draft. Wrong!

The andiron legs should be covered with a good bed of ashes about an inch over the tops of the legs to form a base to build your fire on.

Start by placing a large log known as the backlog eight to ten inches in diameter, or two smaller logs, one on top of the other, against the back of the fireplace, well bedded in the ash. These logs keep the fire from burning out the back bricks and their faces will glow red with coals and reflect the heat of the fire out into the room. It is not intended for them to blaze to any extent.

The fore log, about four inches in diameter, is then placed, bedded down in the ashes in the front of the hearth just in the rear of the andirons. Now the fire, starting with paper and kindling, is built in the center between the back and fore logs. As the fire goes along, larger logs are added. Indeed, logs as large as eight to ten inches can be burned, lasting many hours. It is, however, necessary to build up a hot bed of coals.

Raymond Dyer was an architect from Massachusetts who was interested in homes which utilize alternate energy.

This method of fire burning keeps the draft from passing directly under the logs and results in a slow, glowing fire of small flame and hot coals, needing little attention. As the back or fore logs burn through, they should be moved into the fire and replaced carefully with new ones.

A good fire requires good, dry, hard wood. Oak, maple, birch, hickory, or apple should be cut and dried for six months. Elm and black cherry burn slowly but need a full bed of hot coals to do well.

Should the fire get too hot, it can be dampened with a few shovels of ashes from the side of the hearth. In fact, at night when a fire is wanted the next day, the remaining fire can be fully covered with ashes. In the morning, rake the ashes back and a good bed of red coals will be formed, upon which a new fire can be laid. What's more, the fireplace bricks are still throwing heat during the night.

Maybe we should be thankful for the scarcity of fuel, which will make us find the enjoyment and utility of the fireplace.

A Fireplace Is for Cooking

•

J. C. Lyon

Editor's Note: Fireplaces, notorious for their inefficiency, (more heat usually goes up and out the flue than into the room), can be made more useful if, while the burning logs are producing some heat and giving the room a rosy glow, dinner is cooking away on the hearth. The warmth and beauty of the fire is not sacrificed, and the energy-consuming range need not be used.

We've been cooking in fireplaces over 30 years, ever since we first broiled a steak over the coals in a Greenwich Village apartment. Now, in the mountains of northern New Mexico, three fireplaces supply us with much of our heat, and we cook in all of them, even in the bedroom. We cook roasts, make soups and stews, and bake a little of everything, including pies. (Yeast breads are the exception; except for one that cooks in an open skillet, we haven't yet tackled those.)

No, it isn't as easy as cooking with gas. But if people keep on as we're going, there isn't going to be any gas, for it is a finite resource. Wood, on the other hand, is replenishable. Trees can be farmed—planted, harvested, planted again. Those of us who use wood-burning fireplaces as a source of heat might as well make the most of them and use them for cooking as well.

Most of the needed implements you already have in your kitchen. The others are available. As basics I suggest:

—For roasts, breads, cakes, pies, and casseroles: a large cast-iron Dutch oven with a bail and a raised lip around the lid.

Jetta Lyon is author of The Fireplace Cook Book.

—For soups and stews: a large, enameled pot with its own tight-fitting lid.

—For broiling meats, fish, poultry: a sheet of expanded metal about 12 by 18 inches. (Be sure to scrub it well before using, to remove industrial oils.)

—Firebricks to support the grill or the pot. Ordinary bricks will crack with the heat. Use the hard yellow firebricks from the lumberyard. Two will do; four are better.

—A small oven thermometer to use in the Dutch oven.

—An ovenproof casserole that fits inside the Dutch oven. Also cake and piepans that fit.

—Kitchen tongs.

—Long-handled spoon, spatula, and fork.

—A small brush, with a handle, for dusting off ashes.

—Padded oven mitts.

—Lots of aluminum foil. (If you use foil carefully, smooth and wash and re-use it, you can get a lot of mileage from a single roll.)

Another useful item is a long-handled, two-piece wire grill that folds up, clamping the meat securely inside. For some reason, these are hard to find nowadays. But if you come on one in a hardware store or a camping supply place, snap it up. Ours is 20 years old and still in constant use.

Most of us have broiled a steak or hamburgers over a campfire or a charcoal grill. It's just as easily done in the fireplace. Let the fire burn down to a good bed of coals; set the wire rack across two firebricks; rake the coals under it; put on the steak (or the chicken or hamburgers or whatever) and let it cook.

You'll want to watch it, of course. If it's cooking too fast, rake back the coals; if not fast enough, rake in some more or lower the rack. Time it much as you would under an oven broiler, and when you think it should be done, slide it onto a platter (heated on the hearth) and cut into it. If it needs more cooking, put it back on the rack for a few minutes.

We cook steak quickly over hot coals to keep it rare. Hamburgers can cook faster or slower. To keep chicken tender inside a crisp brown crust, we cook it slowly, a good distance above the coals. I marinate anything that's to be broiled. A good marinade adds flavor, tenderizes, and helps keep foods moist over the open fire.

Along with broiled food we often have baked potatoes. We wrap them in a double layer of heavy foil and bake at the edge of the coals or directly on top. They have to be turned frequently. Depending on the size, potatoes take from one to one and one-half hours. Sweet potatoes take a bit longer. Corn (buttered, left in the husk, and wrapped in foil) takes from 15 to 20 minutes.

For other vegetables, such as peas, broccoli, carrots, green beans, we use small aluminum or stainless-steel loaf pans wrapped in a layer of foil. These sit on the firebricks near the coals and simmer till done. Or if you're using the rack, simply put a saucepan on it and cook the vegetable as you would over a burner. And there's no reason why you can't use a pressure cooker on the rack.

The easiest thing over a wood fire is long, slow kettle-cooking. Once you've put the ingredients together, a stew or a pot of soup can sit on the firebricks and simmer for hours with very little tending.

We do some of our kettle-cooking in two stages. Since we normally have a fire only in the evening, we start a dish one night and finish it the next. Bean soup, for instance. I put the beans in water in my big enamel pot, with a ham bone and some seasoning, set the pot across two firebricks, and let it boil slowly as long as we sit around the fire. The same fire may be broiling that night's supper. Next night, we add some chopped celery, onion, lemon, tomatoes, more liquid if needed, and set it back by the coals. By the time we're ready for supper, so is the soup.

Baking is somewhat trickier. But after a little trial and error, you'll get the hang of it. The main thing is that it takes more watching. I use the big Dutch oven with a baking pan inside. Unless it's a casserole with its own lid, I line the oven lid with aluminum foil. The foil reflects the heat down onto whatever is baking and helps it brown. It also prevents what might be an odd flavor from the condensation of steam on the iron lid.

A trivet or a layer of crumpled foil in the bottom of the oven helps keep food from burning. Put your thermometer in the oven, and while the flames are still high, set the oven by the fire to heat. When it reaches the proper temperature, put in your baking pan, put the lid on, and heap coals on the lid.

Generally, anything in the Dutch oven takes about as long as it would in a regular oven. But there is no way to say *exactly* how long because the heat is neither as even nor as constant. You'll want to turn the oven frequently. And every 10 or 15 minutes, brush the coals off the lid and look inside to see what's going on. If, for instance, a cake is rising too slowly, put more coals on the lid and move the oven closer to the heat. If it's browning too fast, leave the coals off.

Baking anything—pastries, breads, vegetables—does take some practice. Fire is a gift of the gods, and you may have to make a few burnt offerings before you can make it serve you. But I think you'll find it worth the effort.

The Economics of a Wood Cookstove

•

Bonnie Fisher

With the purchase of our farm and old farmhouse in rural West Virginia came an antique wood cookstove. For six months each year our lives revolve around this old beauty, as she not only helps keep us warm and cooks our food, but replaces many of our energy-consuming gadgets, helping us minimize our electric and heating fuel bills.

When we first moved to our farm, we fully intended to heat as the previous owners had, using the wood cookstove to warm the kitchen, and coal stoves in the living room and office. However, we ex-city folk just couldn't get used to waking up in the morning to an icy house. Nor were we prepared for the frozen pipes and wilted house plants we encountered whenever we left the house for a few days in wintertime. After the cat's milk froze in her bowl one night, we decided to install central heating. Then, when the cost of fuel oil soared a couple years ago, we fired up the old wood stove again—and really discovered its many wonders for the first time.

After some experimentation, we have worked out a system so that when the temperature outside dips below 50 degrees F., we combine the efforts of our furnace and wood stove to keep us warm and happy while using a minimum of energy. To keep the temperature of the house constant, we set the thermostat at 65 degrees F. During the day we fire up the stove to heat the kitchen and adjoining living room and office. Late in the evening, when the fire in the stove is down to ashes, the furnace takes over till breakfast. To keep the heat in, we have in-

Bonnie Fisher contributed to the RODALE HERB BOOK, *and runs an herb business in Hickory Hollow, West Virginia.*

stalled insulation throughout the house and replaced our drafty old windows with insulating thermopane.

Many of our neighbors cook with wood exclusively, and claim that food tastes better cooked that way. Throughout the winter I use ours for all my cooking and baking. The vast stove-top area accommodates a wide selection of pots and pans. Temperature control is regulated by distance from the firebox, rather than by push button or knob. Occasionally I retrieve a slightly burned pie or loaf of bread from the oven, but overall my efforts are successful. Putting my electric stove in retirement six months of the year makes a giant dent in our electric bill and energy consumption.

Firewood is abundant on our homestead. Our pastures and woods are full of dead and fallen trees. Instead of piling and burning it, we chop all usable wood into stove-size pieces. We heat our home and clean up the litter on our farm at the same time. With my husband chopping wood and me collecting and stacking logs in the woodshed, both of us get an excellent exercise session.

The stove serves us and saves energy in so many ways. The water tank, for instance, provides plenty of hot water for baths, dishwashing, and various cleaning jobs.

And who needs an electric toaster when crisp toast is easily made by placing bread on an oven shelf for several minutes? (Of course, it doesn't pop out when ready, so a watchful eye is needed to keep it from burning.)

Our wood stove makes a first-rate food dryer. The large surface, oven, warming shelf, and area under the stove all provide the warm, dry conditions needed for speedy dehydration of foods. We like dried foods for their unique flavor. Fruits become chewy and sweet, and oven-dried corn and leather britches have tastes all their own.

When using the oven to dry food, I fill two stainless-steel trays. The door is left ajar and the contents are carefully watched. The shelf and area above the water tank provide ideal conditions for quick-drying of herbs, greens, and beans of all types. The stove also has a dryer on top that operates on the principle of a double boiler. Apples dry nicely in several hours, while greens, herbs, and corn take even less time. Sunflowers, popcorn, and field corn, I dry slowly under the stove. Although I fill hundreds of canning jars and a five-foot freezer yearly, I still like drying best for many fruits and vegetables. No worry about canning lids or freezer shutdowns with our dried foods.

Dave makes excellent yogurt without electricity. He finds the warming shelf just right for the 110 degrees F. required for good yogurt.

I don't believe I've turned on my hair dryer since we've been on the farm. Sitting in my chair by the stove with a good book is the most enjoyable hair dryer I can think of.

Rather than using the clothes dryer for only a few items, I place them on racks in front of the stove. Good woolens and lingerie dry quickly without being subjected to shrinkage and damage.

When dinner is ready and nobody is here to eat it, both the back of the stove and shelf make good food warmers. The frozen meat that I forgot to thaw earlier in the day thaws quickly from the warmth of the wood stove.

In early spring, newborn livestock are often brought in beside the warmth of the wood stove. Newly hatched chicks fallen from their nests find the world isn't such a bad place after all as they soak up the stove's warmth, as do twin ewes who lost their mother. Our newly hatched geese find a warm home in a large box near the hot water tank.

I'm a great collector of iron pots and pans which must be dried immediately after washing to prevent rust formation. After washing, these iron utensils are placed on top the wood stove to dry.

In our part of the country, power outages are common. We can count on several days without electricity each winter. Last winter while the power in our entire county was out for three days we stayed warm and had our regular hot meals. While some of our neighbors stayed in bed for the duration, life on our farm went on as usual.

Our cats also expound the virtues of a wood stove. Where is there a cozier place to nap than under a toasty-warm stove?

After working outside at building fences or cutting wood, there's nothing more satisfying than coming into a warm kitchen to revitalize cold hands and feet in front of a wood stove while sipping a cup of hot mulled cider fresh from the stove top.

In the early spring, the wood ashes—an excellent source of potash and phosphorus—are put to work in the vegetable garden and around ornamentals. Since wood ashes are alkaline, they are highly valued in our garden to help correct our acid soil. We also use them to dust or concoct natural pesticides for the plants.

Such a treasure! If I could keep only one piece of furniture, it would surely be my reliable, economical wood stove.

Pedal Power For The Home

•

Daniel Wallace

One of the main reasons for the success of bicycle-powered machines is the efficiency of human legs in doing work. (see "Power from the People," earlier in Part I). The legs are the strongest parts of the body, and are capable of doing the most work. Legs on bicycles are 95 percent efficient, meaning that the amount of work accomplished by leg action is high in comparison to the effort expended. Such information was mildly interesting to date, considering that the bicycle in this country has (mostly) been a toy for pleasant summer excursions. However, the situation has changed. Today, various efforts are being made by people to exploit that marvelous efficiency ratio of the bicycle to do work.

Homestead Industries of Los Angeles have developed a simple way to convert pedal energy into electricity, but putting a bicycle on a treadmill device which charges a 12-volt battery. This battery can be used in a number of ways, to power small appliances, such as radios, TVs, and record players, or to power light fixtures. The Biko-Generator, as it is called, can also recharge a car battery, if the owner is not inclined to rely upon the bicycle itself.

Other types of bicycle machines have been constructed to perform various home tasks usually done commercially, or by electric appliances. Several homesteaders have modified bicycles to grind wheat and other grains into flour. Experiments with bicycle grain mills at the Rodale Research and Development Center found that the bicycle mill could perform the task on the average of two-thirds less time.

Related to this was that work could be performed for a longer period of time. Hand grinding was usually limited to 12 minute periods, after which the operator needed rest. The bicycle mill finished the same amount of work in four minutes, and the operators were able to continue for various lengths of time, adding up to the exertion equivalent to a vigorous bike ride.

The apparatus developed at the Research and Development Center borrowed from several prototypes in order to come up with a superior machine. The results were a light, but heavy-duty, device, made of 1¼" pipe, with 18" × 30" dimensions. A seat placing the rider's legs close to horizontal made for an easier revolution. A large sprocket was used for more power, and a five pulley system was installed with an idler to reduce slack on changes to different-sized pulleys. The five pulleys allowed the rider to adjust speed as opposed to power, depending upon the task at hand. Hand grips were added for difficult work, and a flywheel smoothed out rough grinding due to changes in rates of pedaling. Finally, the whole machine had a complete set of accessories for different jobs.

The jobs which the bicycle mill is capable of doing range from grinding grain to chopping food, shelling corn, and even to plucking feathers. The important work, however, is that it saves energy by doing these chores instead of machines, in a pollution-free fashion, while adding to the good health of the worker.

Uses for Pedal Equipment:

Food Chopper and Grinder	Sausage/Hamburger Grinder
Food Shredder	Oatmeal Roller
Can Opener	Fish Skinner
Bread Dough Kneader	Rice Polisher
Batter Beater	Grain Cleaner
Ice-Cream Freezer	Butter Churn
Cherry Seeder	Gristmill
Apple Peeler	Corn Sheller
Potato Peeler	Coffee Mill
Knife Sharpener	Kraut Cutter
General Grinder (Tool Grinder)	Battery Charger
Flour Sifter	Air Compressor
Food Grinder	Feather Plucker
Juicer	
Sweet Corn Kernel Cutter	
Irrigation Water Pump	
Meat Slicer	
Meat and Bone Separator, (based on the Paoli design)	
Bone Saw, (like a jigsaw)	
Burrmill and other kinds of Flour Grinders	

Dress Naturally To Meet The Energy Crisis

•

John Feltman

There's a practical, non-polluting way of coping with the energy crunches and lowered thermostats in winter that's simple yet effective: Take advantage of your body's own "heat engine."

Think of your body's natural warmth as a heating element that can be put to effective use in creating a comfortable environment, provided the heat isn't permitted to radiate away into the cold surrounding air. By changing the way you dress indoors and increasing the insulation on your bed, you'll be surprised at just how easily you'll be able to tolerate household temperatures once considered frigid.

The Chinese are well aware of this principle. There's never been enough fuel to squander in China, so the people get along very well in cool homes and public buildings by wearing sweaters both on their upper and lower bodies. They also wear cotton underwear and a heavy cotton suit over the sweater. They are dressed for comfort, if not for style.

One side benefit of keeping your own home cooler in the winter could be liberating yourself from the common belief that the only place to be warm and comfortable during cold weather is indoors. If your home temperature is kept closer to that outside—you won't feel as great a shock when you venture out.

The most efficient, lightweight insulating material we can wear is natural, not synthetic. It's soft, warm goose down. Sewn between layers of fabric in clothing and bedding, goose down conserves plenty of body warmth even in subzero conditions without adding much weight. Dur-

John Feltman is the Managing Editor of Prevention *magazine.*

ing an Alaskan expedition on Mt. McKinley, tests showed lightweight goose-down garments reduced fatigue 40 percent over bulkier synthetic fibers.

A single ounce of goose down contains 23 thousand individual down clusters and more than two million fluffy filaments that interlock and overlap to form a layer of nonconducting still air that keeps warmth in and cold out. Technically, this is referred to as "loft," and it is how thermal efficiency is measured. As goose down "breathes," it transfers body moisture to the outside of the fabric where it evaporates.

Goose down is also super-resilient. A garment can be compressed into a third of its size, but unrolled, it springs back to its original thickness.

If you spend a lot of time outdoors and want to start with basics, goose-down underwear is available. Down vests give you warmth where you need it most without the bulkiness of a sweater. And some down parkas can keep you reasonably comfortable outdoors even when the mercury dips to minus fifty degrees F.

Indoors, it will pay to give special attention to keeping your feet warm. Many people can tolerate a cooler general temperature in their homes but complain of cold drafts around the floor. Down-filled booties are an easy way to solve that problem.

With a down comforter on your bed, you can keep plenty warm overnight even if room temperatures drop into the 50s. Unlike an electric blanket, there's no energy drain and no addition to the electric bill.

If you have a homestead with geese, you might look into the possibility of supplying your own goose down. Many sporting goods stores and backpacking suppliers sell down-filled products.

An Organic Lawn Saves Energy and Cuts Costs

•

Gene Logsdon

Chemicals can make your lawn look as smooth and flawless as a carpet—but if prices continue to rise, that kind of yard may cost as much as a carpet, too. The price of enough lawn fertilizer and weed killer to cover 5,000 square feet climbed to around $17 in 1974. The fertilizer industry says that because of shortages, the price has no where to go but higher. That means that the annual fertilizer and herbicide bill *alone* on a half-acre lawn will be close to $80.

If, however, you think grass ought to look like grass, you can save quite a little chunk of money over a lifetime by managing your yard organically. The secret is to make use of the natural symbiotic relationship between bluegrass and white clover, both of which are practically native plants in most lawns. White clover, being a legume, draws nitrogen from the air and "fixes" or stores it in the soil where the bluegrass is then able to use it for fertilizer.

Louis Bromfield, the noted author, farmer, and conservationist, was one of the first to describe how this grass-legume partnership works. He observed that in renovating poor pastures he had but to apply lime to get a fairly good improvement. The lime made the white clover grow vigorously, storing nitrogen in the soil. When the soil was rich enough, the bluegrass responded to the nitrogen and grew vigorously, dominating the white clover until the nitrogen was used up. Then the clover came back strong, built up the nitrogen reserves, and the cycle started over again.

It will in your lawn too—and though you will still have a few weeds in the process, an occasional mowing will keep your grounds looking very nice indeed. If you spray the weeds with a herbicide, you kill the white clover, destroy the ecological balance, and lock yourself into the garden-store routine of annual chemical replenishment. My former

139

garden-store operator and advisor always used to tell me that "the best herbicide is a rich soil because it keeps the grass growing vigorously enough to crowd out most of the weeds automatically."

The bluegrass-white clover love affair works best when soil pH is between 6.5 and 7.0. About two pounds of lime per 100 square feet ought to do it every three years. If you over-lime, you get too much white clover in relation to bluegrass. If your soil has been acid (below 6 pH), liming may cause white clover to appear in your lawn you didn't know was there.

If you do need to seed white clover, do so at a rate of ½ pound per 1,000 square feet. Broadcast it on bare land or on your sod in late winter or very early spring some morning when the ground is frozen. The tiny seeds fall into the pitted, frozen surface, and when the ground thaws, the soil flows together enough to cover the seed.

Bluegrass should be sown on tilled soil in August when there's sufficient moisture present, or in the spring after the ground dries out and warms up a little. You can throw the seed on bare ground and it will usually germinate, but you'll get a better stand by covering the seed and firming the soil over it.

When you mow, don't remove all the clippings if your soil is on the poor side. The clippings eventually rot into humus and provide the replenishing nutrients your grass needs along with the nitrogen from the clover. If you remove all clippings, then eventually you'll have to fertilize with something.

Unfortunately, too many people have been brainwashed into believing that white clover is a weed. But the legume has many merits in the lawn. It will catch on poor land and resist drought well. It is low growing, and practically indestructible. The low, white, sweet-smelling blossoms hold a beauty all their own. What's more, at today's fertilizer prices, clover can store well over $100 worth of nitrogen in an acre of lawn every year, without any expenditure of your money or of fossil fuel energy. And that's kind of beautiful, too.

Feeding the Family: Best Ways to Prepare for Food Security

•

Jeff Cox

The magic formula to avoid being drowned by the coming social tidal wave is: *get involved in your own future and maximize your independence,*" write Paul and Anne Ehrlich in "The End of Affluence."

These are tumultuous times, and prudence tells us that *this* year's garden should be pointing toward self-sufficiency in the kitchen:

—Every bite of food that we grow ourselves frees farm production for the help of the millions who face starvation around the world.

—Organically grown food goes easy on the natural resources (like energy) that are spiralling upward in price, aggravating the social turmoil.

—Homegrown food is downright *cheap* compared to supermarket food. With the right program, it can be virtually free.

While some look on food independence as a means of security in unstable times, growing your own is probably the best thing we can personally do to *ease* the new future into being. It is preventive, positive, and humane—and doesn't have to be a knee-jerk reaction from fear.

Your food independence program can be as elaborate or as simple as you wish. The Indians of North America grew little besides corn, beans, and squash—which happen to be a nutritionally excellent trio.

Both the corn and beans have good stores of protein, points out Frances Moore Lappé in "Diet for a Small Planet." However, when eaten in *combination*, top-quality protein is unlocked for the body's use that would be unusable if they were eaten separately. The squash contributes some good protein, but even more importantly, it's chock full of vitamins and minerals—especially vitamins A and C.

An area of food security you should not forget about is the raising of small livestock. During World War II, with meat and eggs rationed or

141

unavailable, many people raised rabbits and chickens in addition to tending a garden.

The advantages of raising small animals are many, but economics is probably of prime importance. Rabbit meat presently sells for almost two dollars a pound, but can be produced at home for less than one-fifth the price. Goat milk, if you can find it, goes for nearly one dollar a quart, but when produced at home runs around ten cents a quart.

Most small stock can be raised as a part-time hobby. You certainly don't have to be a full-time farmer to have a few chickens, geese, rabbits, ducks, pigeons, sheep, and goats around the place, although it does require a fair amount of space.

One highly productive item you should plan on including that doesn't require very much work is a beehive. Pesticides are killing off America's bee population at an alarming rate of almost two percent a year. After the first year, a hive can be counted on for at least 50 pounds of honey a year. Another benefit you will receive from keeping bees is greatly improved pollination in your garden, resulting in increased yields.

Foods to consider for a food independence program should be able to be stored without electricity or power. Esther Dickey, in "Passport to Survival," urges families to store whole wheat berries, salt, honey, and powdered non-instant milk. All these store for long periods of time in a dry attic.

Any leafy or green vegetable can be shredded and dried. That includes Swiss chard, broccoli, cabbage, Chinese cabbage, lettuce, parsley, peppers, and more. People who are drying their vegetables should be sure to blanch their vegetables before drying to stop destructive enzyme action.

You can allow a certain amount of your spring peas to dry on the vine and harvest them for easy storage in a dry place. Your pumpkin patch and stand of sunflowers will give you bushels of the highest-quality nutrition for pennies in those crops' seeds—which are easily stored.

Nuts from wild nut trees or your own nut trees are another high-quality food that keeps well in the attic. Potatoes and onions need no mechanical devices to keep well all winter. Carrots, beets, and turnips keep in a root cellar or even in a barrel sunk into the ground. Cabbage, properly treated, turns itself into sauerkraut—nutritious and needing no power to store. Power-packed parsley dries nicely. Kale will grow just about all winter in the garden. Fruit can be turned into delicious leather (a chewy form of dried fruit), apples can be schnitzed into slices and dried peaches can be sun-dried for storage. And you should take the trouble to know the wild foods called lambs-quarters, nettles, and pigweed—all are very nutritious and grow over a wide range. Nettles are 50 percent protein.

As mentioned above, aim for storing food without power. Just about any vegetable can be stored dry—and the ones that can't, like tomatoes, can be put up in mason jars. "In our farmhouse, food is stored in every nook and cranny, from attic to cellar," says Ariel Wilcox in "Food Storage on the Maine Farmstead" in *Farmstead Magazine*. A good book on food storage is what's needed here, and "Stocking Up," by the editors of *Organic Gardening and Farming*, edited by Carol Stoner, covers all the necessary information.

Growing all your own food doesn't have to be a full-time occupation. It can still be the family hobby. And it will save the family a big chunk of change.

When we asked our friend Ruth Stout, who's in her nineties, to estimate the yearly expenses and savings from her garden, she replied, "Last year my garden expense was $35—feeding two people all year. If you mean by 'value,' what would I pay if I bought the vegetables from a store, how could I know? I haven't been inside a supermarket for over 12 years and I never buy a vegetable." If Ruth does it at 90, that should be encouragement for us.

John McMahon of Clifford, Inc., says he is heading toward full subsistence on the one to two acres he cultivates. He figures his seed cost at about $50 a year. His animals provide all the fertilizer he needs. "My initial equipment expenditures were made over a decade ago. Replacement, repair, and maintenance are all we have now—that's $25 or $30 a year," he says. About 80 percent of the food for the four people who eat at the McMahon table is raised right on the farmstead, and he says the food is worth "$1,000-plus . . . maybe $2,000 now."

Ona Raney Shope takes about $200 worth of food from her garden—and that goes a long way toward feeding her (she's 75) and her new husband of 18 months (he's 72).

We polled 20 organic gardeners to find their garden expenses, the value of the food they grow, how many are in their family and the percentage of their food that they grow. The results showed that these organic gardeners are well on their way to self-sufficiency. They are already doing what world leaders are calling for the developed nations to do: Simplify their eating habits and go lightly on the land so that famine will pass the world by like a lion that looks but moves on without pouncing.

What's it take to feed a family for a year from the home garden? Here's a list of 16 common vegetables you'll most likely grow as staples for winter storage. It includes a rough estimate of how much to store for each person, approximate yields per 100 foot row, distance between plants, and how much seed you'll need to plant 100 feet. From these estimates, decide how much your family will need to get through the non-growing season, how much space in your garden it will take, and how much seed you'll need. Remember these are rough averages, so plan on planting more, and include enough for you to enjoy during the growing season. You'll probably also want to plant seasonal crops for eating fresh from the garden—vegetables such as zucchini, lettuce, radishes, spinach, okra and many others.

Name	How Much To Store Per Person	Distance Between Plants	Yield Per 100 Ft. Row	Seed Needed For 100 Ft. Row
Beans (bush)	18 quarts	4-6 inches	50 quarts	12 ozs.
Beets	75 roots	3 inches	300 roots	1 oz.
Broccoli	15 heads	18-24 inches	60 heads	2 pkts.
Brussels Sprouts	8 quarts	18 inches	90 quarts	1 pkt.
Cabbage	20 heads	18 inches	70-90 heads	2 pkts.
Carrots	100 roots	3 inches	400 roots	½ oz.
Cauliflower	9 heads	18-24 inches	70 heads	1 pkt.
Corn	100 ears	12-18 inches	65 ears	1½ ozs.
Cucumbers	75 pickles	36-60 inches	300 fruits	2 pkts.
Onions	115 bulbs	2-3 bulbs	300 bulbs	2 lbs. of sets or ½ oz. seed
Peas	25 quarts	6 inches	60 quarts	1 lb.
Peppers	50 fruits	12-18 inches	400 fruits	4 pkts.
Potatoes	75 pounds	12-15 inches	100 pounds	8 lbs. seed potatoes
Tomatoes	25 quarts	24-48 inches	100 quarts	24 plants
Turnips	35 roots	3 inches	300 roots	2 pkts.
Winter Squash	15 fruits	48-120 inches	50-60 fruits	½ oz.

Here are the results of the survey of the 20 organic gardeners:

Average Cost-Value of 20 Organic Gardens*			
Yearly garden expenses $54	Value of food grown $538	No. in family 3–4 people	Percentage of food grown 60 percent

*The gardens ranged from $100-value kitchen gardens to $2,000 whoppers that feed a large family and more. Eliminating the gardens over $1,000 leaves an average food saving of about $300.

Index